Studying in the Content Areas

The Sciences
Second Edition

By Carole Bogue, Ph. D.
Canada College
Redwood City, California

H&H Publishing Co., Inc.
Clearwater, Florida

H&H Publishing Company, Inc.

1231 Kapp Drive
Clearwater, FL 34625
Ph (813) 442-7760

Studying In the Content Areas
The Sciences
Second Edition
By Carole Bogue, Ph.D.

Editing, Production, and Interior Design:
Karen H. Davis

Editorial Assistant:
Priscilla Trimmier

Production Supervision:
Robert D. Hackworth

Business Operations:
Mike Ealy, Sally Marston

ISBN 0-943202-38-8

Library of Congress Card Catalog Number 93-079042

Printing is the lowest number: 10 9 8 7 6 5 4 3 2

PREFACE

STUDYING IN THE CONTENT AREAS: THE SCIENCES carefully teaches students study strategies and leads students to apply key study skills to material in science content areas. The major difference between this book and other study skills textbooks is that it contains actual material drawn from this area for completing the exercises. As a result, students are provided the opportunity to practice the application of the skills using relevant material. The topics addressed in the sciences material are common to those treated in introductory college courses in these areas. Many study skills textbooks inform students what they need to know about study skills and what is involved in their applications, but they do not provide ample opportunity for practice ensuring mastery of all study skills, including taking notes from lectures and taking tests, nor do many texts pay special attention to particular content areas. A unique feature of this book is the accompanying audiocassette which contains lectures treating science topics. This addition provides students the opportunity to practice taking notes from lectures in the sciences. Furthermore, in addition to the material being drawn from the sciences, special tips related to course expectations in the sciences are provided. The ultimate goal of this textbook is to enable students to complete college coursework in the sciences successfully.

This text may be used in both lecture classes and in individualized lab programs. A sufficient number of features permit effective utilization in both settings.

The textbook begins with a section on applying study strategies, followed by a section divided into nine skill areas, as well as Sample Chapters from science material. These chapters are to be used in completing the skill exercises. In each section covering all basic study skills, the following parts are included in the order listed:

1. An **overview** of each skill instructing students in the application of the study skills to the given content area.
2. **Exercises** providing opportunity for practice of each skill in the various subareas of the content areas addressed.
3. A **Summary** of the skill.
4. **Sample Chapters** used to complete most study skill exercises.

An AUDIOCASSETTE contains lectures from the sciences. This audiocassette is used when students are ready to apply Skill 7: Taking Notes from Lectures.

ACKNOWLEDGEMENTS

I am indebted to the hundreds of students at San Jose City College who worked through all of the skill exercises written for the various drafts of this textbook and to the students and faculty at Canada College who worked through the Applying Study Strategies material. Their participation in this project provided much insight and reaffirmation of the final direction taken.

Appreciation is also expressed to Bob King, San Jose City College, who served as a Consultant for the sciences, and to Lois Janowski, San Jose City College, Research Assistant, who contributed significantly in areas of research, editing, and assembling the material into its final form.

CONTENTS ████████████████████

INTRODUCTION

Purposes of Book

STUDYING IN THE CONTENT AREAS: THE SCIENCES

There are many reasons for taking science courses. You may want to prepare yourself for a career that requires a background in this area. Or maybe you want to know more about this area before you make a career choice. Perhaps, for your own information, you simply want to explore the topics treated in physics, chemistry, and/or biology.

Whatever reasons you have for studying this academic area, good study skills and strategies can help you get the most out of your courses. The purposes of this book are the following:

1. To describe for you what typical science courses will be like and how to succeed in such courses;

2. To provide you with exercises so that you can practice applying study skills and strategies in this content area;

3. To review with you how to apply study skills when you study this academic area; and

4. To provide you with additional suggestions for succeeding in science coursework.

Performance Objective

A performance objective tells you what strategies and skills you will learn and be able to use. When you have completed this book, you will be able to apply the following study strategies when completing courses in college:

1. Attitude
2. Motivation
3. Time Management
4. Anxiety Levels
5. Concentration
6. Information Processing Skills
7. Selecting Main Ideas
8. Study Aids
9. Self Testing
10. Testing Strategies

You will be able to apply the following study skills when you study the sciences.

1. Surveying textbooks
2. Surveying textbook chapters
3. Building vocabulary
4. Reading and marking textbook chapters
5. Using maps, diagrams, graphs, and tables
6. Taking notes from reading assignments
7. Taking notes from lectures
8. Taking tests
9. Using the library

How to Use This Book

This book is divided into three sections: 1) Applying Study Strategies, 2) Applying Study Skills with nine skill areas, and 3) Sample Chapters for the Sciences. Each strategy and skill area consists of the following parts in the order listed:

1. An **overview** of each skill/strategy showing how it is applied to the content area.

2. **Exercises** providing the opportunity for practicing each strategy and skill, the latter using material in each subarea of a given content area.

3. A **Summary** of the strategy/skill.

Sample Chapters of science material are used when completing most of the study skill exercises.

For Skill 7, **Taking Notes from Lectures,** a tape on which lectures in each subarea have been recorded is provided for completing exercises after the Skill 7 explanation and Summary have been studied.

For **study strategy exercises,** check your responses with your instructor. Since every college student's set of circumstances and reason(s) for attending college may be different, there is no absolute right or wrong answer for the items in these exercises; consequently, an answer key for strategy exercises is not provided. All students, however, need to know and use the strategies discussed.

An answer key for all **study skill exercises** is found in the back of the book.

1. For each study skill, read the section, complete the **Exercises,** and review the **Summary.** To complete the exercises for some of the skills, use the **Sample Chapters** found behind the exercises. Page numbers for the **Sample Chapters** are provided at the top of the exercises.

2. To complete exercises for Skill 7, **Taking Notes from Lectures,** use the tape which accompanies this book.

3. After completing each skill exercise, check your answers with the answer key found at the back of the book. Ask your instructor about any topics that are still not clear to you. You may repeat or review any part of each section whenever you need to do so. And, of course, you may ask your instructor for help whenever you need it. When you have successfully completed work for the study skills covered in the content area, you will be ready to take the test for that section. Before the test, read over the skill **Summaries** to make sure you understand how to apply the skills. Look over the exercises you have completed. Then you will be better prepared to take the test.

Applying Study Strategies

There are many reasons for taking courses in college. You may want to earn a degree, prepare for or advance yourself in a career, explore subject areas for your own information and/or enjoyment, and/or develop skills useful in many aspects of life.

Whatever reasons you may have for attending college, good study strategies can help you succeed in college.

APPLYING STUDY STRATEGIES leads you, the student, to apply the habits essential to success in college. The major difference between this section and the study skills section of this book is that the focus here is placed on STRATEGIES rather than skills. Strategies may be described as those habits which college students need to improve to be successful in college study.

Characteristics of a College Student

There are some characteristics of any college student that **cannot** be changed. A number of typical examples can be found in the individual's history:

- ☞ A college student's actual birth date cannot be changed.
- ☞ A college student's experience learning to read cannot be changed.
- ☞ A college student's experience learning English in the 9th grade cannot be changed.
- ☞ A college student's high school grades cannot be changed.
- ☞ A college student's experiences at home and in school as a child cannot be changed.

Some characteristics of college students are quite desirable and **should not** be changed. Some examples might be:

- ☞ A college student with a positive attitude about college would want to keep that attitude.
- ☞ A college student with a valuable scholarship would not want to lose it.
- ☞ A college student with good physical health does not want his/her health to change.
- ☞ A college student with high motivation would not want to change that motivation level.

Other characteristics of college students, however, **can and should** be changed. Some examples might be:

- ☞ A college student's financial situation may be too close to poverty. Some improvement may be most desirable if academic success is to be attained.
- ☞ A college student's eyesight may be hindering his/her reading ability. Glasses or contact lenses may make the difference between success and failure in school.
- ☞ A college student, during his/her past experience in school, may have developed certain attitudes or habits which will greatly interfere with future academic success. Changing those attitudes or habits may significantly improve academic performance.
- ☞ A college student's learning strategies may not be effective. Learning strategies can be changed. Poor ones can be eliminated and good ones can be improved.

This book is concerned only with those aspects of a college student that can be changed or improved. The intent is to provide you with information that can be used in improving your ability to eliminate poor habits and learn new ones that will lead directly to increased levels of academic performance. Nothing suggested in the remaining pages focuses on that which cannot be changed. Every suggestion can be implemented; doing so will take some effort and practice, but anyone can improve his/her learning strategies.

Before reading the next subsection, further introducing you to learning strategies, see if you can list the following:

    a.   three facts from your individual history that cannot be changed but that may influence your performance in college;

    b.   three of your positive characteristics that should not be changed;

    c.   three of your characteristics that could and should be changed in order to improve your academic performance.

Course-Specific Knowledge

Whether you are taking a mathematics test, writing an essay, or giving an oral report in a history course, the quality of the result is dependent on some course-specific knowledge.

No one can hope to perform successfully in school without course-specific knowledge. But that knowledge is not sufficient to guarantee academic achievement. Some students devote much time and effort to attempts to learn the knowledge, but nevertheless continue to have real difficulty. The students who place all their emphasis on course knowledge overlook two other, very important, aspects of their learning.

Desire to Learn

A second important aspect of a student's success is the desire to do well. The student who wants to succeed in school is far more likely to perform better than the student without that desire. This comment, while obvious, is not meant as a criticism of those lacking desire to learn. For some students there are good reasons why desire to succeed in school is low; it is suppressed by other interests and concerns. Financial problems and personal interests outside of school may well compete with the desire to achieve academic success. A student who wants to succeed on the football field, on the dance floor, or in the theater may have extremely high motivation toward such activities but those desires may conflict with school success.

Even the student who has a strong desire to learn as well as course-specific knowledge is not guaranteed academic success. Knowledge and desire are not sufficient; another aspect of learning is also needed.

Management Ability

A third aspect of learning is the ability to manage both course-specific knowledge and desire to learn. Without good management skills, knowledge and desire are insufficient for achieving academic success. For example, many people know all the facts about the dangers of cigarettes and they have a sincere, strong desire to stop smoking. Nevertheless, they continue smoking. Why? They lack the management skills required for implementing the steps they should follow to quit the habit.

The student who possesses course-specific knowledge, desire to learn, and management skills can compete well in all academic tasks. To help you better understand your own role in becoming a better student, the remainder of this section presents ten learning strategies. Each of these learning strategies is important by itself and each can be studied and developed separately. All, in combination, are needed for academic success.

Ten Learning Strategies

1. Attitude
2. Motivation
3. Time Management
4. Anxiety Levels
5. Concentration
6. Information Processing Skills
7. Selecting Main Ideas
8. Study Aids
9. Self-Testing
10. Test Strategies

In conclusion to this introductory section, see if you can now create several lists as follows:

   a.   your performance levels in three essential skill areas: reading, writing, and math;

   b.   three academic subjects you expect to enjoy most (for each, briefly describe the source of your expectation);

   c.   three major goals you now have (for each, briefly describe the relationship between your goal and success in college);

   d.   three outside areas of interest that could compete with your desire to succeed in college coursework;

   e.   three types of management decisions you need to make constantly when you are studying; and

   f.   three learning strategies you think you should improve.

Strategy 1
Attitude

In order to succeed in college, you need to have a positive attitude, closely linked to setting direction and reaching desirable, realistic long-range goals in life. Given a clear set of goals, you can determine ways to reach them.

APPLYING THIS STRATEGY

To apply this strategy, take notice of your attitude as you read the next few pages. Mentally alert yourself to the effect of your attitude on both your interest level and your memory.

> ## PERFORMANCE OBJECTIVES
>
> When you have completed Strategy 1, you will be able to approach the task of setting clear goals. You will be able to specify some long and short term goals in the major areas of your life.

Attitude

Attitude refers to a learning strategy that drastically affects your desire to do well in school and complete the coursework assigned. Attitude might as easily be termed "Setting Long-Range Goals." Attitude deals with attaining control of your life, setting long-term direction, and planning realistic objectives. Long-term direction and related objectives could include working in a certain field or at a certain level within an organization, but they may be more specific. For example, one student might be committed to becoming a pediatrician and working at Memorial Hospital while another student, with equally valid goals, might be committed to working with people in third world countries in an attempt to improve their health care.

Many students enter college with attitudes reflecting unclear long-term goals. These students may not have made final decisions, or their own decisions, regarding the direction their lives should take. Perhaps they have postponed final decisions or perhaps someone else has imposed choices upon them. Clarifying long-term goals is essential for these students if they are to get the most out of college.

Two lists of attributes regarding attitude are shown on the next page. Each item illustrates some attribute of a student whose attitude helps or hinders his/her academic performance. A blank space appears before each item. If that item seems to fit your situation, place a check mark (✔) in the blank.

Likely Attributes Demonstrating a Helpful Attitude

____ Focuses on a clearly stated major for college
____ Studies well independently and undertakes individual projects
____ Concentrates on academic tasks for reasonable periods of time
____ Understands the relationship between college and the outside world

Likely Attributes Demonstrating a Harmful Attitude

____ Shows impatience while completing tedious assignments and preparing for tests
____ Desires to socialize in lieu of keeping up with coursework
____ Possesses an unclear notion of long-range goals
____ Possesses a vague idea of how college will lead to long-range plans

At some times in their lives, even the best students may have been able to honestly check any of the last four examples. If you checked them, there is no need to panic. You now have some clues to help you start planning to improve your attitude.

A major step toward improving attitude is making a list of your plans and dreams for the future. Writing down possible goals will help you clarify and recall your intentions. Further, it will assist you in determining if your goals are realistic. Finally, writing down possible goals will help you determine ways to reach them. Ultimately, placing goals in writing helps you see the relationship between success in college and your specified long-range goals. A clear understanding of this relationship will have a positive effect on your interest, diligence, and memory when taking courses in college. With increased interest in your courses, for example, you will be able to improve concentration, motivation, and use of time; you will be able to achieve greater academic success.

Setting goals requires that you consider various aspects of your future. Although it may be difficult to know exactly where you hope to be five or ten years from now, you should give consideration to important areas including education, career, social life/family, and financial status (income, savings, investments, property). Remember, the goals you set should represent your interests and what you really want in life. Such goals will be exciting to you and will be easier to attain than goals which represent general societal expectations.

Review: Attitude

Let's review what you need to know and do in order to set goals.

1. Consider at least four major areas of life such as education, career, social life/family and financial status. Think about long term goals (five to ten years from now) for each area.

2. A long term educational goal may be to complete a BA/BS degree in a chosen field. Perhaps it requires other degrees. It definitely looks further ahead than one semester or one year.

3. In terms of career, a long term goal may be to attain a full-time position as a high school science teacher, to work as a researcher in a government research lab, or to gain employment as an allied health worker. It could include some summer work in preparation for a permanent position, but it must not be limited to such short-term employment.

4. When considering social life/family, you should know that it's important to live a balanced life and to include time for family, friends and recreational activities, even when you're taking a full load of difficult courses. Without friends and/or family, loneliness and feelings of isolation may develop which can work against your ability to succeed in college. Too much time spent with friends and family, however, may interfere with time needed for study and impede success in college.

   A long term goal in this area of life might be to have a successful marriage and two children.

5. In terms of financial status, it is important to analyze current needs and future income and investments. A long term goal might be to earn $40,000 plus annually and to have a well-balanced investment portfolio with a portion of your savings invested in stocks, a certain percent in bonds, some funds in CDs, and the balance in money market funds.

EXERCISES ▰▰▰▰▰▰▰▰▰▰▰▰▰▰

There are three exercises for Attitude. To complete the first two exercises, you
will need to write goals in the chart provided and respond to questions about long-
range goals.

EXERCISE 1 ▰▰▰▰▰▰▰▰▰▰▰▰▰

Think about the goals you would like to reach within the next five to ten years.
Write your goals in phrases in the spaces provided.

LONG TERM GOALS 5 – 10 years from now	
Education: (courses/certificates/ degrees)	
Career: (field of work/ level of position)	
Social Life/Family (dating relationships/ marriage/friendships)	
Financial Status: (income/savings/ investments/property)	

EXERCISE 2 ▰▰▰▰▰▰▰▰▰▰▰▰▰▰▰▰▰▰▰▰

Respond to the questions about long term goals. Check Y for Yes, N for No, or N/A for Not Applicable.

Y	N	N/A	
			1. Are your long-range goals attainable?
			2. Is the career for which you are preparing suited to your interests and abilities?
			3. Have you visited organizations relevant to your career choice and observed jobs involved in this field?
			4. Have you contacted others working in this field and discussed the kinds of jobs involved?
			5. Have you worked in or involved yourself in any way with the career of your choice?
			6. Have you considered lifestyle as an important factor embedded in all long-range planning?

EXERCISE 3 ▰▰▰▰▰▰▰▰▰▰▰▰▰▰▰▰▰▰▰▰

✎ Writing is one of the best ways to clarify your thoughts about Attitude. Turn to page 300 and write a short passage about your attitude.

Other Activities for Improving Attitude

At your college library, locate several autobiographies written by famous or well-known people. Read four or five interesting selections, but only those portions of the books which discuss adolescence or early adulthood. You are likely to find some famous people who seemed linked from birth to a plan for the future. You are also likely to find some famous people who struggled greatly with planning their lives. Reading about others will help you begin to plan for your future.

In addition, talk with adults whom you respect. Explain that you want their perspective on how they chose a life's plan. Such conversations can be of great interest and can provide you ideas regarding how to establish long-range goals that are suited to you.

SUMMARY ▬▬▬▬▬▬▬▬▬▬▬

A good attitude for completing college courses successfully is linked to attaining control of your life, setting long-term goals, and planning realistic objectives. In order to improve your attitude for attaining success in college, certain steps can be taken as follows:

1. Consider the major areas of life including education, career, social life/family and financial status.

2. In each area, specify long range goals based on personal interests and abilities and other considerations specific to your own circumstances.

3. Place your goals in writing. Putting your goals in writing helps clarify them. Later, it helps you remember them and evaluate their attainment. Finally, it helps you see the relationship between college work and your specified long range goals.

Strategy 2
Motivation

A high level of motivation to learn and complete assignments in a timely manner is essential to success in college. Motivation is related to short-term goals for successful completion of the tasks in courses in which you are enrolled.

APPLYING THIS STRATEGY

To apply this strategy, think now of your reason for reading this material on motivation. Maybe your only reason is that the reading is a required assignment. It is to your advantage to have other, more personal reasons. As you read the next few pages, reflect on your reasons for doing so.

PERFORMANCE OBJECTIVES

When you have completed Strategy 2, you will be able to discuss motivation and take certain steps which lead to increased motivation.

Motivation

Motivation is another learning strategy that drastically affects your desire to manage your time and do well in college. Whereas attitude is associated with long-range goals, motivation deals with short-term goals. Motivation affects the immediate or near-term tasks necessary for successful completion of coursework. Your motivation is reflected in the way that you handle your course responsibilities.

The essence of motivation is the acceptance of personal responsibility for whatever happens to you. Students with high motivation give themselves credit for all their achievements. Such students even credit themselves for lucky events, usually reflected in such statements as, "Being in the right place at the right time." These students credit themselves with being there at the time. Students with low motivation are more likely to blame much of what happens to them on forces beyond their control. Claims such as, "The teacher was terrible," "I just can't do that type of work," and "My alarm didn't go off," are expressions symptomatic of poor motivation.

Two lists of attributes are shown below. Each item in the lists illustrates some attribute of a student whose motivation helps or hinders his/her academic performance. A blank space appears before each item. If that item seems to fit your situation, place a check mark (✓) in the blank.

Likely Attributes Demonstrating High Motivation

___ Always attends class and participates in discussions
___ Works on assignments far in advance of their due dates
___ Studies as diligently for courses disliked as for favorite courses
___ Makes a connection between short-term tasks and long-range goals
___ Manages time well and devotes more than sufficient time to completion of assignments and test preparation

Likely Attributes Demonstrating Low Motivation

___ Makes excuses for lack of achievement
___ Crams for tests
___ Skims assigned reading
___ Spends much time socializing

At some times in their lives even highly motivated students may have been able to honestly check any of the last four examples. If you checked them, there is no need to become discouraged. You now have some clues regarding how to increase your motivation.

Review: Motivation

Let's review what you need to know and do in order to increase your motivation.

1. Initially, as stated earlier, you need to accept responsibility for all of your school tasks. Rather than blaming others or outside circumstances for your performance, you need to determine exactly what you need to do in order to succeed and take those steps promptly.

2. In order to accept responsibility for school work and increase your motivation, in general, you should begin to act motivated. The student who acts motivated (even if he/she isn't) finds that school and specific courses become easier and more interesting. For example, motivated students complete reading assignments before class. They arrive at every class on time and participate in class discussions. They ask questions about concepts not clearly understood. After class, they review their notes, complete assignments, and seek help if necessary before the next class. Further, motivated students work on long-range assignments far in advance of their due dates. They review coursework regularly and prepare routinely for all quizzes and tests. Finally, they give some thought to the relationship between successful course completion and long-range goals. As a consequence, motivated students and those who act motivated are more in control of their school life and accept greater responsibility for all tasks. Like a snowball rolling down a mountain, a small positive change in motivation can grow and grow.

3. A first step in acting motivated is to take inventory of time spent on outside class assignments. It is helpful to make a list of all specific assignments due in the next month. Beside each assignment, write an estimate of the time required to complete it properly. This information may be a revelation to you because it may indicate why it is impossible to do everything well at the last minute. Then track the number of hours you do spend as you complete the assignments listed. If the actual number of hours is far less than your estimated hours, you may need to evaluate the quality of your work. If the actual number of hours is greater than what you estimated, you may need to reevaluate your estimates and how you schedule your time.

4. Finally, in order to accept responsibility for all coursework and to increase motivation, it is important to analyze your outside interests and responsibilities and determine if they are interfering with your academic performance. Outside interests may interfere with motivation to reach short-term goals related to academic success as well as desire to learn. Outside activities likely involve your family, your economic situation, and your social life. Responsibilities in these areas are important but they may also greatly interfere with your school motivation. It is impossible to remain

highly motivated when outside activities require so much time that little is left for class attendance and completion of assignments. Something always suffers. When lack of motivation is caused by too many responsibilities, it can only be solved by the careful elimination of some of those responsibilities.

EXERCISES ▰▰▰▰▰▰▰▰▰▰▰▰▰▰▰▰

There are three exercises for Motivation. To complete the first exercise, you will need to write course assignments for a 30-day period in the chart provided. To complete the second exercise, you will need to respond to questions as indicated.

EXERCISE 1 ▰▰▰▰▰▰▰▰▰▰▰▰▰▰▰▰

Complete the following chart by entering the name of each course in which you are enrolled and any related reading assignments, tests, and other assignments which have been given for the next 30-day period. Indicate the amount of time you think you will need to complete each task. Use this chart to check against the actual time you spend completing these tasks.

Thirty-Day Period

Course	Reading Assignments	Time Needed	Tests	Time Needed	Other Work	Time Needed

Total Hours _____   Total Hours _____   Total Hours _____

Grand Total _____

EXERCISE 2 ▰▰▰▰▰▰▰▰▰▰▰▰▰▰▰▰▰

The following questions address outside responsibilities in areas of family, economics, and social life. Check Y for Yes, N for No, or N/A for Not Applicable. Determine whether or not you need to reduce outside activities.

Y	N	N/A	
			A. Do family activities interfere with your scheduled study time?
			B. If necessary, is there any way that you can reduce the number of social/family activities or reschedule them so you won't compromise time reserved for study?
			C. Is the time you spend on social activities appropriate, given the demands of college?
			D. Can you maintain your friendships and/or relationships if you must reduce the number of social activities which you typically schedule?
			E. Are your finances stable and sufficient enough to meet your basic needs while in college?
			F. Does your work schedule interfere with time needed to study and attend classes?
			G. Do you work too many hours and suffer fatigue?
			H. Can you change your work schedule if it hinders success in college?
			I. Have you determined a way to reserve some funds for emergencies while you are in college?

EXERCISE 3 ▰▰▰▰▰▰▰▰▰▰▰▰▰▰▰▰▰

✎ Writing is one of the best ways to clarify your thoughts about Motivation.
Turn to page 302 and write a short passage about your motivation.

SUMMARY ▬▬▬▬▬▬▬▬▬

Motivation is associated with short-term goals or successful completion of course-work. The essence of motivation is the acceptance of personal responsibility for all courses and related tasks. In order to increase your motivation, certain steps can be taken as follows:

1. Determine exactly what you need to do in order to succeed in college and take those steps promptly.

2. Begin to act like you are highly motivated. Follow the steps which highly motivated students take while in college.

3. Estimate the time needed for completion of outside class assignments and test preparation. Compare the total with the actual hours spent on these tasks. If great discrepancies are found between your estimated and actual hours, look closely at the quality of your work and the time you spend completing tasks.

4. Analyze outside interests and responsibilities related to family, economics and social life. If such activities are so numerous and time consuming that they compromise your successful completion of coursework, consider ways to reduce them.

Strategy 3
Time Management

Successful college students manage their time well and maintain a high level of organization with regard to daily routines. Time management requires creating realistic schedules for the term, week, and day.

APPLYING THIS STRATEGY

To apply this strategy, think now of the time you have reserved to complete this reading. Did you plan in advance to do this reading now? Is this reading part of an overall schedule for completing your work? Reflect on these questions as you read the next few pages.

PERFORMANCE OBJECTIVE

When you have completed Strategy 3, you will be able to
organize your daily routines by creating a time schedule,
maintaining a weekly appointment book, and completing a
daily "TO DO" list.

Time Management

Time management is a learning strategy that is primarily a management skill.
Because most students have a variety of demands placed on their time, there is a
tendency for them to do whatever task immediately confronts them rather than to
create a realistic schedule and adhere to it. Time management is the making of
schedules, daily and weekly, that prioritize activities to be accomplished. Further,
time management requires the discipline to follow schedules created; too many
students allow themselves to be drawn away from scheduled tasks.

Creating schedules in order to manage time well requires a certain degree of
circumspection. You must know yourself, how you learn best, the best time of
day to schedule certain tasks, and your personal likes and dislikes. This knowledge and self awareness facilitate creation of realistic schedules which can be
implemented successfully.

Two lists of attributes are shown below. Each item illustrates some attribute of a
student whose time management helps or hinders his/her academic performance.
A blank space appears before each example. If an example seems to fit your
situation, place a check mark (✓) in the blank.

Likely Attributes Demonstrating Successful Time Management

___ Knows when mathematics will be studied this week
___ Goes to the library as scheduled to prepare a term paper
___ Allows adequate time for social life
___ Joins his/her study group, as scheduled

Likely Attributes Demonstrating Unsuccessful Time Management

___ Says "Yes" when invited to have a cup of coffee although this activity
deviates from the time scheduled to study math
___ Wonders when to study for an upcoming test
___ Questions whether to read the assigned book or see a movie about the topic
___ Wonders why his/her roommate keeps a desk calendar for daily "TO DO"
activities

At some time in their lives even the best students may have checked any of the last four examples. If you checked them, there is no need to drop out of school. You now have some clues regarding how to improve your time management.

Review: Time Management ▰▰▰▰▰

Let's review what you need to know in order to manage your time well.

1. To become proficient as a time manager, you need to prepare and manage a schedule. Within that schedule, you need to block out sufficient time for study. Usually three-hour blocks of time are recommended, even though it may be necessary to take a break every fifty minutes or so. More than three hours could well exceed limitations of concentration and energy. In other words, two 3-hour blocks of time scheduled on two consecutive days are better than one 6-hour block scheduled on one day. Also, when deciding how to arrange your schedule, remember the Carnegie rule: For every hour of class, you should commit to two hours of homework/study. Lab courses require less work at home but more time in the lab on campus; the same formula though, three hours per week for each unit of credit, applies to term length lab courses. Depending on the course and instructor, the Carnegie "rule of thumb" varies, but it serves as an excellent guideline when arranging a schedule. You should complete a time schedule at the start of each term; it will serve as a basis for your weekly schedule.

   Included in a time schedule should be your class schedule, time at work if you have a job, travel time, time for sleep, meals, recreation, and time reserved for study. A sample time schedule is shown on the facing page.

TIME SCHEDULE

	Sun.	Mon.	Tues.	Wed.	Thurs.	Fri.	Sat.
6:00	sleep	sleep	sleep	sleep	sleep	sleep	sleep
7:00	sleep	dress/eat	dress/eat	dress/eat	dress/eat	dress/eat	sleep
8:00	sleep	travel to sch	travel to sch	travel to sch	travel to sch	travel to sch	sleep
9:00	eat	Hist 100	Eng 100	Hist 100	Eng 100	Hist 100	eat
10:00	study	study		study		study	family
11:00	study	Psych 50	Math 80	Psych 50	Math 80	Psych 50	family
12:00	study	eat	eat	eat	eat	eat	study
1:00	Recrea-	job on	study	job on	study	job on	study
2:00	tion with	campus	study	campus	study	campus	study
3:00	family	↓	study		study	↓	recreation
4:00		travel	travel	↓	review	↓	
5:00		recreation	recreation	study	recreation	travel	
6:00		eat	eat	study	↓	relax with	↓
7:00	↓	study	study	travel	eat	friends	dinner
8:00	eat	study	study	eat	study		with
9:00	relax	study	relax with	relax with	study		family
10:00	sleep	sleep	friends	friends	sleep	↓	
11:00	sleep	sleep	sleep	sleep	sleep	sleep	sleep

2. Once a basic time schedule is completed, you need to keep a weekly appointment book. In a weekly schedule, you may incorporate your class schedule, but memorize your class schedule soon after the start of each term. It is crucial in your Weekly Appointment Book to note due dates for assignments and special projects, appointments with instructors, field trips planned, test dates, and special study hours scheduled prior to tests or exams. On the next page is a sample page from a Weekly Appointment Book.

WEEKLY APPOINTMENT BOOK

MONDAY	THURSDAY
	11 am, Math Quiz
7pm, meet John to study for history test	

TUESDAY	FRIDAY
	10 am, Review for Psych quiz
1-2 pm, work in math lab	

WEDNESDAY	SATURDAY
9 am, History test	
12 noon, see Psych instructor	*1 pm, Library to prepare term-paper for history*

	SUNDAY
	10 am - 1 pm, Library, term-paper for history

3. As you review your weekly appointment book, you need to keep a separate daily "TO DO" List. This will include in priority order all tasks that must be accomplished during the day and unexpected activities as well. You should make a "TO DO" List every day, use it as a guide, and check off items as you complete them. Keep this list in your Weekly Appointment Book. Also, use it when you develop your list for the following day just in case some items couldn't be completed and need to be added to the next day's list. In your daily "TO DO" List, routine items like eating, travel and the like need not be included. Below is a sample page from a daily "TO DO" List.

DAILY TO DO LIST   DATE: _Friday, 9/10_

1. *Drop clothes at cleaners on way to class.*

2. *Review for Psych Quiz from 10:00 to 11:00 a.m.*

3. *Check oil in car.*

4. *Stop at grocery store for study snacks.*

5. *Purchase pack of note cards to do library research tomorrow.*

EXERCISES

There are three exercises for Time Management. To complete these exercises you will need to complete a Time Schedule, Weekly Appointment Book page, and a "TO DO" List.

EXERCISE 1

Complete the blank Time Schedule below indicating your current classes, work hours, sleep, travel/eating time, time reserved for others, recreational activities, and study time.

	Sun.	Mon.	Tues.	Wed.	Thurs.	Fri.	Sat.
TIME SCHEDULE							
6:00							
7:00							
8:00							
9:00							
10:00							
11:00							
12:00							
1:00							
2:00							
3:00							
4:00							
5:00							
6:00							
7:00							
8:00							
9:00							
10:00							
11:00							

EXERCISE 2 ██

Complete the Weekly Appointment Book page below. Insert possible due dates
for assignments/special projects, appointments, tests /exams, study group meetings, and any other special events which might be planned.

WEEKLY APPOINTMENT BOOK

MONDAY	THURSDAY
TUESDAY	FRIDAY
WEDNESDAY	SATURDAY
	SUNDAY

EXERCISE 3 ▰▰▰▰▰▰▰▰▰▰▰▰▰▰▰▰

Complete a sample Daily "TO DO" List focusing on one of the days of the week in the Weekly Appointment Book completed for Exercise 2. Write some possible tasks in the blanks provided.

DAILY TO DO LIST   DATE _____

1.

_____

2.

_____

3.

_____

4.

_____

5.

SUMMARY ▰▰▰▰▰▰▰▰▰▰▰▰▰▰▰▰

Time management is crucial to meeting the numerous demands placed on you while in college. Time management involves creating a realistic schedule enabling you to prioritize tasks for each term and to refine this general prioritization with weekly and daily schedules. Creating time schedules is of little value, however, unless you discipline yourself to follow them.

In summary, the following steps need to be taken to schedule your time:

1. Create a Time Schedule at the start of each term to include time for courses, work, sleep, travel, eating, recreational activities, and study.

2. Maintain a Weekly Appointment Book to include due dates for assignments/special projects, appointments, tests/exams, study group meetings, and other special events.

3. Keep a Daily "TO DO" List for each day of the week. Include all special tasks that must be given priority that day.

Strategy 4
Anxiety Levels

In order to succeed in college, you need to take and retain control of your schedule and course completion. You need to apply methods to lower anxiety levels because anxiety reduces the control you have over your ability to learn.

APPLYING THIS STRATEGY

To apply this strategy, begin this reading with some awareness of your comfort level. Is your body tense? Is your mind at ease? Reflect on the questions as you read the next few pages.

PERFORMANCE OBJECTIVES

When you have completed Strategy 4, you will be able to apply specific methods designed to reduce high anxiety levels.

Anxiety Levels

Anxiety is a psychological/physiological state that is entirely normal. However, some students experience such high levels of anxiety that it greatly interferes with their ability to learn. An important learning strategy for all students is to develop methods for lowering anxiety in those situations where it is sabotaging all other learning. Techniques for lowering anxiety are part of your management skills. When high levels of anxiety are reduced, both your desire to learn and ability to acquire knowledge will increase. Sometimes, successful efforts to reduce anxiety are followed by dramatic improvements in academic achievement.

High levels of anxiety are generally accompanied by both psychological and physical distress. The student suffering from anxiety may experience fear, hate, intimidation, and worry that seem completely inappropriate to the situation. The student may also suffer severe sweating, heavy breathing, stomach problems, and tenseness. The source of these physical reactions is anxiety.

Because the effects can be so debilitating, students with high anxiety make little or no progress in learning. They are distracted by the anxiety. Their normal ability to concentrate is greatly decreased. They seem to be listening, but do not hear. They are so fearful that they avoid the source of their fear whether it be a subject, a teacher, a classroom, or even a book. When anxiety is a serious problem, the student often engages in extremely negative self-talk that impedes learning, motivation, long-range plans, and opportunity to make friends and maintain relationships.

Two lists of attributes are shown below. Each item illustrates some attribute of a student whose anxiety level helps or hinders his/her academic performance. A blank space appears before each example. If that example seems to fit your situation, place a check mark (✓) in the blank.

Likely Attributes Demonstrating Low Level of Anxiety

___ Feels confident in his/her ability to be successful with academic tasks
___ Feels in control in academic situations
___ Seeks out professors for help or discussion without hesitation
___ Keeps a weekly study schedule that varies with particular class assignments

Likely Attributes Demonstrating High Level of Anxiety

___ Believes that there are some academic tasks that he/she cannot complete
___ Feels uncomfortable with class activities requiring student participation
___ Procrastinates when some academic tasks are assigned until all others have been completed
___ Experiences undesirable physical symptoms when completing certain academic tasks

High levels of anxiety greatly interfere with academic success because they distract students from the tasks needing time and effort. The student with high levels of anxiety in mathematics, for example, often neglects mathematics assignments and/or misses math classes more often than other classes. A student with high levels of test anxiety spends precious time during the test worrying about some lack of knowledge or chastising her/himself for being a "stupid" person who didn't learn the material. Further, such high anxiety can lead students to "draw a blank" and forget the concepts and information they really do know.

Review: Low Anxiety Levels ▮▮▮▮▮▮▮▮▮▮▮

Let's review what you need to know in order to reduce your anxiety level.

1. Most anxiety is accompanied by an associated lack of control which may or may not be generalized to all courses. The student suffering anxiety with mathematics may only experience a lack of control when in a mathematics classroom or while taking a mathematics test. Such a student may not suffer anxiety in any other settings. Another student may suffer anxiety for all courses in which he/she is enrolled during a term. All students suffering anxiety need to find, perhaps with the assistance of the instructor, methods for regaining control. These methods may focus on test taking, working on assignments, and/or participating in class activities. Students suffering anxiety should discover options available for reducing anxiety levels in academic settings.

2. An excellent approach for most students suffering anxiety is "overstudying." The student needs to consciously take steps that might be completely unnecessary with other activities: sit at the front of the classroom; arrange frequent conferences with instructors; do each assignment early; do it over again; prepare for every test thoroughly; read all recommended materials; study with a friend; visit the Learning Center with a request for further help; and, most importantly, persist, persist, persist.

3. Physical symptoms which may accompany anxiety need to be relieved because any decrease in physical symptoms will lead to a direct decrease in the psychological discomfort. A simple but effective method for diminishing physical symptoms relies on paying attention to your breathing. Don't try to change your breathing; just pay attention to it. A normal result of attending is the lowering of the physical symptoms. Practicing this process controls the physical symptoms of anxiety. You should practice paying attention to your breathing when you are calm. Doing so will help you apply this process when you are under stress.

4. For students with severe anxiety, there are programs and materials aimed at reducing that anxiety. Some of these materials are listed in "Other Activities for Lowering Anxiety Levels." Instructors and counselors may suggest other materials and activities if the student will lower his/her anxiety level enough to ask for help.

Remember, anxiety is a problem because it takes time and effort away from the essential tasks that must be completed. Finding ways to lower anxiety levels gives you extra hours every day, extra days every week, and extra weeks every year.

EXERCISES ▰▰▰▰▰▰▰▰▰▰▰▰

There are two exercises for Anxiety. To complete the first exercise, respond to the items in the spaces provided. Write your responses in complete sentences.

EXERCISE 1 ▰▰▰▰▰▰▰▰▰▰▰▰▰▰▰▰

1. Describe how you would "overstudy" if you suffered high anxiety for a course you were taking.

2. How would you relieve physical symptoms of anxiety if you were highly anxious about a particular course you were taking?

3. Describe a situation in which you have suffered anxiety regarding a particular course or subject.

4. What recommendation would you give to a friend to reduce his/her anxiety for mathematics?

5. How does high anxiety reduce the opportunity to achieve academic success?

EXERCISE 2 ▰▰▰▰▰▰▰▰▰▰▰▰▰▰▰▰

✎ Writing is one of the best ways to clarify your thoughts about Anxiety. Turn to page 304 and write a short passage about your anxiety.

Other Activities for Lowering Anxiety Levels

1. Read the following books which may be found in your college library:
    a. Herbert Benson, **The Relaxation Response**, William Morrow and Co., Inc., New York, 1975
    b. Robert Hackworth, **Math Anxiety Reduction**, 2nd Edition, H&H Publishing Co., Inc., Clearwater, FL, 1992
    c. Martin Seligman, **Helplessness**, W. H. Freeman and Co., San Francisco, 1975
2. Ask someone in your college's Learning Center for materials related to lowering anxiety levels.
3. Check for special seminars or workshops on lowering anxiety. Your campus probably offers such activities during each term.

4. In order to reduce stress/anxiety levels regarding college in general, you should learn about your college of choice and discover what resources are available there to assist you. Learning about your college will help you adjust to college from the start. Consider the following:

    a. Attain a copy of the college catalog and examine it.

    b. Attain a copy of each schedule of classes as soon as it's available; use the schedule to select courses for the following term.

    c. Read the student handbook. It contains information about special opportunities available to you and outlines student expectations.

5. Visit the student service areas and instructional resources available on campus. Most students need to use resources to increase their chances for success in college.

SUMMARY

Anxiety reduces the ability to concentrate and complete assignments, attend class meetings and participate in class successfully, prepare for and take tests, and in general, to make progress in a day-by-day manner toward successful course completion.

In order to reduce anxiety level, several steps can be taken:

1. Ask your instructor about methods for regaining control. Methods for reducing anxiety may focus on taking tests, completing assignments, and/or participating in class activities. Student-selected options in these areas are available in many academic settings.

2. "Overstudy." Take every possible step to attain success when taking a course causing anxiety.

3. Attend to your breathing. Practice when you feel calm and apply the process when you are feeling anxious. This attending method helps you control the physical symptoms of anxiety.

4. Read books treating the topic of anxiety and seek suggestions from counselors and instructors for materials and activities designed to reduce anxiety.

Strategy 5
Concentration

The ability to concentrate is the ability to focus attention on a particular task or assignment without being distracted. An ability to concentrate, associated with a desire to succeed at a task, is required for academic success.

APPLYING THIS STRATEGY

To apply this strategy, take notice of anything that distracts you for the next few minutes. Are there sounds or people that take your attention away from the reading? Does your mind wander, reducing your ability to recall the words and the meaning of a passage? Reflect on these questions as you read the next few pages.

PERFORMANCE OBJECTIVE

When you have completed Strategy 5, you will be able to take steps which lead to improved concentration.

Concentration

As studied in the last section, high levels of anxiety can disrupt concentration. Anxiety is only an example of the many emotional distractions that every student may encounter. Other emotional distractions such as anger or love can also impede concentration. The most common emotional distraction for most college students is found in their social lives. The desire to maintain and strengthen social relationships is often in direct conflict with the need to concentrate. A recognition of this conflict is a first step in avoiding the type of emotional distractions that most frequently interfere with concentration while studying.

Physical distractions present in every environment can also reduce concentration. Noises from a radio, a nearby conversation, an air conditioning unit, or a passing airplane may interfere with concentration. Temperature which is too cold or too warm has a harmful effect on concentrated study as well. Lighting which is too dim, too bright, or too varied may negatively affect concentration. Furniture in the study environment may have a negative effect on the comfort level which affects concentration. Reading in bed may be a good idea for approaching sleep, but a terrible distraction when preparing for a test. Students need to be aware of such distractions and control them whenever possible.

In addition to recognizing emotional and physical distractions, students also need to know that good concentration is facilitated by application of key study skills during scheduled study hours. Application of these skills, treated in the second section of this book, helps you focus on the material and improve concentration during lectures, labs, and study hours.

Two lists of attributes are shown below. Each item illustrates some attribute of a student whose ability to concentrate helps or hinders his/her academic performance. A blank space appears before each example. If that example seems to fit your situation, place a check mark (✔) in the blank.

Likely Attributes Demonstrating Good Concentration

___ Recognizes and avoids distractions that interfere with concentration
___ Takes clear, well-organized notes from reading assignments and lectures
___ Carefully marks articles assigned for a course
___ Maintains a well-planned, carefully arranged study area
___ Demonstrates an interest in her/his studies that supercedes other distractions

Likely Attributes Demonstrating Poor Concentration

___ Rarely takes notes from reading assignments
___ Takes disorganized and incomplete lecture notes
___ Studies in a variety of places
___ Spends more time making social contacts than studying
___ Displays fidgety, restless physical activity while studying

Everyone suffers from lack of concentration. There will be times, situations, and social settings when concentration will suffer. A wise student learns to recognize those periods when concentration is lacking and reshapes or controls the situations so concentration will be possible. Only with adequate concentration can effective study occur.

Review: Concentration

Let's review what you need to know in order to improve concentration.

1. Attitude (long-range goals) improvement, increased motivation (short-range goals), time management skills, and methods for lowering anxiety will have a desirable effect on concentration. In addition, you need to apply the key study skills described in the next section of this book in order to focus on material heard and read.

2. Recognizing a distraction is a first step towards avoiding it.

3. Controlling your study environment will aid concentration. Good lighting, comfortable temperature, and low noise levels will improve the ability to concentrate.

The suggestions above require you as the learner to be an active rather than passive or uninvolved learner. Active learning means becoming involved with the learning task and taking conscious steps to understand, organize, and retain new information.

EXERCISES

There are two exercises for Concentration. To complete the first exercise, respond to the items on the facing page in the spaces provided. Some items require you to write responses in complete sentences.

EXERCISE 1 ▰▰▰▰▰▰▰▰▰▰▰▰

1. What type of noise or noise level is most conducive to your ability to concentrate? Why?

2. Describe (using complete sentences) a study area that would be conducive to good concentration.

3. Describe (using complete sentences) an active learner.

4. What aspect of life serves as a major source of emotional distraction for many college students? What steps will you take to avoid this problem?

EXERCISE 2 ▰▰▰▰▰▰▰▰▰▰▰▰

✎ Writing is one of the best ways to clarify your thoughts about concentration. Turn to page 306 and write a short passage about your concentration.

SUMMARY ▰▰▰▰▰▰▰▰▰▰▰▰

Improving concentration requires that you take steps recommended for Attitude (long-range goals), Motivation (short-range goals), Time Management, and Low Anxiety Levels.

The more active you become as a learner, the better you will be able to concentrate when studying.

Recognizing distractions and controlling your study environment will lead you to improve your concentration.

Strategy 6
Information
Processing Skills

In order to learn new information efficiently, you need to apply information processing skills. Two major aspects of processing new information include relating it to prior knowledge/experiences and organizing it into meaningful structures depicting interrelationships among main points, supporting details, and minor points.

APPLYING THIS STRATEGY

To apply this strategy, begin by seeing yourself as a wonderful storehouse of information. You know all about your life up to the present moment. Use that knowledge as your read the next few pages. Try to make connections between your reading and your personal experiences.

PERFORMANCE OBJECTIVES

When you have completed Strategy 6, you will be able to describe information processing skills and take steps which lead you to improve them.

Information Processing

For many students, learning in school is the same as amassing information. Sadly, some instructors encourage students to memorize vast amounts of information rather than to organize new information into understandable wholes and relate it to prior knowledge. Information processing skills enable students to give meaning to separate pieces of information. An ability to process information well is directly related to the acquisition of knowledge.

There are two major aspects of good information processing. The first is the continuous use of prior knowledge. The second is an ongoing attempt to organize new information. These two aspects may be discussed separately, but they actually work in combination.

Every student comes to any new learning situation with prior knowledge that will be of great value if it is accessed and then used. Students have some prior knowledge of many topics, but their knowledge may need to be uncovered as the study begins. Students will also have some prior knowledge of their best approach for learning about a new topic. This prior knowledge may include their experience with earlier topics in the course itself.

In order to gain a clear understanding of new information, students must organize new knowledge. This task requires a conscious recognition of major points, underlying concepts, supporting details, and minor points. An expert in a subject considers a course to contain only a few powerful concepts which explain a myriad of related questions or problems. The novice considers a course to contain a vast number of seemingly unrelated pieces of information which require an inordinate amount of time to learn. Information processing skills need to be used to organize new information into structures of knowledge with interrelationships among different pieces of information clearly understood.

Two lists of attributes are shown below. Each item illustrates some attribute of a student whose ability to use information processing skills helps or hinders his/her academic performance. A blank space appears before each example. If that example seems to fit your situation, place a check mark (✓) in the blank.

Likely Attributes Demonstrating Effective Use of Information Processing Skills
___ Explains new ideas in relation to past experience and/or prior knowledge
___ Organizes main points and supporting details readily
___ Asks questions like, "What is the relationship between this point and that one?"
___ Reviews previous material in a course on a regular basis
___ Discovers relationships among ideas in different courses

Likely Attributes Demonstrating Ineffective Use of Information Processing Skills
___ Believes it is easier to memorize information than to develop an understanding of it and relate it to other information
___ Forgets material learned after the final instead of reviewing it for the next course in the sequence
___ Writes notes as if all ideas or points are parallel instead of organizing them as main points, supporting statements, and minor points
___ Wonders why some courses are required since they seem to be clearly unrelated to career goals

Review: Information Processing ▰▰▰▰

Let's review what you need to know in order to apply information processing skills.

The student seeking to improve information processing skills needs to develop the habits of using prior knowledge and organizing information into a hierarchical arrangement. Each time a new topic is introduced, the user of prior knowledge tries to make some immediate connections between new information and prior knowledge. Sometimes these connections prove to be temporary, but the process, when practiced, always improves understanding and retention.

The difference between raw information and organized knowledge can be compared to the difference between a pile of bricks and a brick wall. No organization is evident in a pile of bricks. In a brick wall, each brick supports the others and in turn is supported by them. Each brick in the wall, connected to the others with mortar, has a position in relation to the other bricks. Bricklaying is a skill and learning to organize information is a skill. If organizing information is lacking from your repertoire, begin now to deal with information as you would when building a brick wall. Look for relationships and insert the connections (mortar) that will hold ideas or points properly. The ideas or points clearly connected and organized are of greater value than separate pieces of disorganized information.

EXERCISES ▰▰▰▰▰▰▰

There is one exercise for Information Processing. To complete this exercise, you will need to respond to the questions below in the spaces provided. Write your responses in complete sentences for Questions 2 and 3.

EXERCISE 1 ▰▰▰▰▰▰▰

1. Draw a structure that would help you describe your family.

2. What prior experience may help a novice who is trying to learn the art of sailing a boat?

3. How does relating new information to prior knowledge and/or experiences help you attain a clearer understanding of the new concepts or ideas?

SUMMARY ▆▆▆▆▆▆▆▆▆▆▆▆▆▆▆▆

Information processing involves relating new information to prior knowledge and/ or experiences and organizing new information into meaningful structures including interrelationships among main ideas, supporting details, and minor points. It's important to develop the habit of relating new information to prior knowledge and experience in order to facilitate the learning and understanding of new information.

Strategy 7
Selecting Main Ideas

When completing academic tasks such as reading assignments, studying textbooks, and listening to lectures, you need to be able to select the main ideas read and heard. Selecting main ideas is a first step in organizing information into meaningful structures which clarify interrelationships among the various aspects of the points discussed.

APPLYING THIS STRATEGY

To apply this strategy, try to find the main idea in each paragraph on the next few pages. After reading each paragraph, underline a sentence or phrase that seems to summarize its meaning.

PERFORMANCE OBJECTIVES

When you have completed Strategy 7, you will be able to specify typical contexts in which you must select main ideas. Also, you will be able to take steps which lead you to select the main idea.

Selecting Main Ideas ███████████████████

Textbooks, lectures, classroom discussions, special projects, and all other instructional activities involve content material that ranges from extremely important to very trivial. The student needs to have a screening device that identifies the most important information, accurately associates major details clarifying main points, and discards the relatively unimportant points.

Some students have acquired a skill in applying the process described above. For those students with skill at selecting main ideas, studying is efficient because it is well directed. Some students have not acquired a skill at selecting main ideas; they make inefficient use of their time. These students often labor long and hard to learn everything in a course without differentiating between the absolutely necessary points and the very minor details. Students with poor skills at selecting main ideas often experience frustration, overwork, and failure.

Main ideas are presented in all lectures, textbooks, and other instructional materials. Most professors do not hide the relative importance of ideas from their students. In fact, a common remark for professors is, "Now, this is very important." Such a remark clearly tells you that a main idea will follow. In textbooks and other instructional material, headlines, boldface print, color, boxes, and other visual clues are aimed at emphasizing the main ideas of a section. Look for these clues when you first preview a textbook and use them as you search for main ideas in the reading material.

Two lists of attributes are shown on the following page. Each item illustrates some attribute of a student whose ability to select main ideas helps or hinders his/her academic performance. A blank space appears before each example. If that example seems to fit your situation, place a check mark (✔) in the blank.

Likely Attributes Demonstrating Effective Selection of Main Ideas

___ Takes well-organized and succinct lecture notes

___ Enters in textbook pages distinct marking for main ideas, major support statements, terms and their definitions, and items in lists

___ Makes accurate predictions of likely test questions

___ Creates summary sheets showing the main points relating to a topic

Likely Attributes Demonstrating Ineffective Selection of Main Ideas

___ Uses magic marker highlights on entire sections within reading assignments

___ Possesses no clues regarding what to study for a test

___ Takes indecipherable or messy lecture notes showing a collection of one-line comments

___ Creates incomplete summary sheets showing little relationship between textbook and lecture notes regarding a topic

Review: Selecting Main Ideas

Let's review what you need to know in order to select the main idea.

1. A complete paragraph always contains a main idea, a statement which indicates who or what the paragraph is about and what is important about that topic. The main idea can be the first sentence in the paragraph, a sentence preceded by introductory comments, a concluding sentence, or implied. When main ideas are implied, the reader is required to read the related, supporting details and specify a statement indicating the main point.

2. In longer selections consisting of several paragraphs, the main idea is a summary statement of the main ideas of the separate paragraphs. In chapter-length textbook sections, clues may point to main ideas: introductory sections, different colored ink used for main points, main points placed in boxes on chapter pages, illustrations showing the relationship of main ideas, and summary sections.

3. Professors often give you clues regarding the main ideas when they use words like:
    Today, I will discuss . . .
    An important feature . . .
    The chief reason . . .
    It's important to know that . . .
    There are five steps to this process as follows . . .
    To summarize . . .

EXERCISES ◼◼◼◼◼◼◼◼◼◼◼◼◼◼

There is one exercise for Selecting Main Ideas. Respond to the questions below
in the spaces provided. Write your responses in complete sentences.

EXERCISE 1 ◼◼◼◼◼◼◼◼◼◼◼◼◼

1.   What does the main idea of a paragraph tell you?

2.   What does the main idea of a longer selection tell you?

3.   During a lecture, what are three clues/signals that main points will follow?

4.   What is a first step in organizing information into meaningful structures?

SUMMARY ◼◼◼◼◼◼◼◼◼◼◼◼◼◼

It is important when completing academic tasks to be able to select the main ideas
presented. Selecting the main idea might begin at the paragraph level. Main
ideas must be articulated for longer selections also. In terms of longer selections,
study aids found in textbooks such as bold print, headings, and boxes often help
you select the main points. During a lecture, professors often provide signals/
clues that main points will follow. Including all main points in your lecture notes,
as well as noting all main points in reading assignments is a first step in organizing information into meaningful structures.

Other Activities for Finding the Main Idea

1.   Complete exercises requiring you to get the main idea; such exercises are
     found in numerous materials prepared for improving reading comprehension.
2.   Take a course to improve reading comprehension skills offered at many
     colleges.

Strategy 8
Study Aids

In order to increase opportunity for success in college, students should use the study aids available to them. Aids are available within textbooks and other required material. Supplementary material also aids students in their studies. Further, students may create their own study aids.

APPLYING THIS STRATEGY

To apply this strategy, think about what the next few pages are likely to contain. Can you list some study aids you already use in your present courses? Are there places on your campus where additional study aids are made available to students? Reflect on these questions as you read the next few pages.

PERFORMANCE OBJECTIVE

Upon completion of Strategy 8, you will be able to discuss three types of study aids which help students learn more efficiently and effectively.

Study Aids

Study aids are valuable features of many instructional materials available to students. Some students use study aids effectively. Other students ignore the study aids that are available; they limit their use of assistance which is available to them.

Several types of study aids are used by successful students. The first type consists of those supplementary materials prepared by someone else for the benefit of students. For example, textbook publishers often create computer software, video tapes, and other materials to accompany their textbooks. These expensive study aids are available on most campuses to any student who wishes to make use of them. In fact, someone on your campus may have coordinated all study aids available. Learning Centers, labs, libraries, and offices of Learning Specialists often make collections of supplementary materials available for student use. In these areas, free tutoring, sample tests, and special study groups may be available too for the student who seeks them.

Another type of study aid consists of those materials that students create for their own use. For example, some students create drawings and charts to illustrate points and organize information. They also create questions about the material and write summaries to help them understand and retain information to be learned. Students often share their aids with others in study groups and, in this way, help each other learn more effectively.

A third type of study aid consists of those features found in instructional materials which help students process the information to be learned. Examples include the table of contents and preface, special headings often in boldfaced print and dark color, main points repeated in rectangles or other figures in the margin, numbered lists included in context, maps, charts, tables, diagrams, chapter summaries, review sections, indexes, and glossaries. Students who are aware of these aids tend to make much better use of their study time.

Two lists of attributes are shown below. Each item illustrates some attribute of a student whose use of study aids helps or hinders his/her academic performance. A blank space appears before each example. If that example seems to fit your situation, place a check mark (✓) in the blank.

Likely Attributes Demonstrating Effective Use of Study Aids

____ Uses the features in the textbook that serve as study aids
____ Visits campus resources where supplementary materials are available to students
____ Creates charts and diagrams as he/she studies
____ Shares self-created study aids with study groups

Likely Attributes Demonstrating Ineffective Use of Study Aids

____ Skips over the study aids in a textbook
____ Seldom takes notes or creates charts/diagrams to illustrate information to be learned
____ Rarely uses supplementary material available on campus
____ Studies alone and rarely joins a study group to discuss material and share student-created aids

The student who makes little use of study aids may find it helpful to look closely at his/her attitude and motivation. Study aids are available to all students. Some effort may be needed to find and make use of them, but the time and effort required serve as a valuable investment.

Review: Study Aids

Let's review what you need to know in order to use study aids effectively.

1. You need to create study aids for your own use in order to be a successful student. Taking notes from reading assignments and lectures leads you to create your own aids including lists, charts, and diagrams. When notes are clearly taken, summaries for review can be prepared easily too.

2. Look for and note special features of textbooks and chapters which serve as excellent study aids. Illustrations such as diagrams, graphs and tables often make ideas or facts clearer and help you retain the information presented. Maps are used to help you picture a place or an area.

3. Many publishers provide supplementary materials to aid you in learning the material in a textbook. Most campuses have areas such as Learning Centers, labs, and libraries where supplementary material is collected and made available to students.

EXERCISES

There is one exercise for Study Aids. Respond to the items on the following page in the spaces provided. Write your responses in complete sentences.

EXERCISE 1

1. What study aids are included in your social science textbook?

2. What study aids are not found in your textbook but would be helpful if they had been included?

3. What are two kinds of study aids that students often create for themselves?

4. What are two kinds of supplementary materials which serve as study aids?

5. On your campus, where can supplementary materials be found?

SUMMARY

Several types of study aids are used by students as follows:

1. Special features of textbooks and textbook chapters
2. Supplementary materials provided by publishing companies
3. Student-created study aids such as diagrams, lists, charts, and summaries

Effective use of study aids will increase your opportunity for academic success. Students who use study aids study both more efficiently and more effectively.

Strategy 9
Self Testing

Self testing, with frequent, regular review, is needed for academic success. Frequent self testing confirms knowledge and understanding of subject matter. It also uncovers areas in which information is missing or not clearly understood.

APPLYING THIS STRATEGY

To apply this strategy, take a moment to reflect on the previous material in this book. Are you using the ideas and suggestions presented earlier? Do you remember the ideas and suggestions? A study skills course is successful only if you begin improving your ability to study. Reflect on these questions as you read the following pages.

PERFORMANCE OBJECTIVE

When you have completed Strategy 9, you will be able to describe the steps you need to take to apply self testing.

Self Testing

Self testing as described here is equivalent to frequent and regular review. The student who makes review a regular part of her/his study activities achieves many benefits from that process. The student who fails to review on a regular, systematic basis tends to be overwhelmed when preparing for an imminent test.

Self testing should be applied during every study session. For example, class notes need to be reviewed after each class meeting. Any notes that are not well understood form the basis for good questions that need to be answered at the next class meeting. The notes can be used as a source of similarities and differences among previous topics in the course. You should test your knowledge and understanding using the notes containing examples and applications of concepts presented. At this time you also should relate new knowledge to previous topics covered. Doing so establishes contact with prior knowledge and helps you make predictions about what may likely occur.

Self testing should also be applied at the completion of certain segments of a course. For example, at the end of each chapter it is probably wise to set aside a study session just to review notes taken from the entire chapter. Some of the same self testing activities used with daily class notes could apply to these more structured reviews. Class notes, reading assignments, problems, and projects for the chapter can be integrated. Key points that are understood can be listed and summarized. Key points that are inadequately understood need to be followed up immediately with your professor or study group.

Periodically, it is helpful to review all notes taken to date. In this way you begin to see the larger picture, and interrelationships of ideas become evident. During these reviews, you need to integrate your lecture notes, reading assignment notes, exercises, problems, and projects.

Two lists of attributes are shown below. Each item illustrates some attribute of a student whose ability to effectively use study aids helps or hinders his/her academic performance. A blank space appears before each example. If that example seems to fit your situation, place a check mark (✔) in the blank.

Likely Attributes Demonstrating Effective Use of Self Testing

___ Reviews lecture notes as soon as possible after each class meeting
___ Identifies what he/she knows and what is not clearly understood
___ Combines lecture and reading assignment notes into understandable summaries
___ Reviews regularly instead of "cramming"

Likely Attributes Demonstrating Ineffective Use of Self Testing

___ Frequently skips studying between two class sessions
___ Spends little time identifying what she/he knows and what she/he doesn't understand clearly
___ Considers each new chapter a separate part of the course
___ Discovers that some class notes make little sense when studying for a midterm or final

Students often postpone self testing until a test is scheduled, when it is too late for full effectiveness. Review and self testing need to become a part of every study session; the sooner they follow instruction the better they facilitate student learning. Review pays great dividends because it integrates old and new learning. Further, it improves understanding of the older material and positively affects ability to remember all material covered.

All learners experience instances when they fail to grasp the material. Those students who regularly review become aware of what they don't know or understand and, in a timely manner, seek other instruction to overcome deficiencies in learning. Those students who do not engage in review risk academic failure. They do not discover what they don't know; they do not have an opportunity to correct errors and overcome their inadequacies. Learning difficulties are correctable, but only when the learner uses good review procedures.

Review: Self Testing

Let's review what you need to know in order to apply self testing.

1. When reviewing notes taken from lectures and reading assignments, it is important to read all notes carefully, supply missing information, and clarify notations that will make little sense later. Do this as soon as possible after the lecture or reading assignment.
2. As you review, check yourself and see if you can recall, without referring to your notes, the important points regarding each main idea included. Recite the material aloud during this review in order to check your recall. Make sure at this time that you understand the concepts presented and the related information. Ask your instructor, study group members, and/or tutor for clarification if understanding is incomplete.
3. During self testing, students will need to refer to all material studied and marked in order to check illustrations and main points discussed. Maps, diagrams, graphs, and tables help you review effectively.
4. When reviewing notes from reading assignments, refer to the original source. Study the illustrations again, complete and/or review any exercises provided, and review the summary at the end. Pay special attention to any new vocabulary words and make certain the meanings of such words are clearly understood.
5. Complete a review or self testing process after each class session and after taking notes from each reading assignment.
7. Periodically, review all notes taken and combine them into understandable outlines, concept maps, or summaries depicting interrelationships of ideas. A more lengthy review should be conducted after completing each textbook chapter, section, or other course segment. Reviews should be conducted at regular intervals throughout the term, too, at which time all notes taken to date are combined and reviewed. These broader reviews might be conducted every few weeks.

EXERCISES ▐███████████████████████

There is one exercise for Self Testing. Respond to the items below in the spaces provided. Write your responses in complete sentences.

EXERCISE 1 ▐███████████████████████

1.   When should you first review class notes and when should you first review notes taken from reading assignments?

2.   When you review material, what is a good way to check your recall?

3.   When should you review all of your notes taken to date from both class sessions and reading assignments?

4.   Why is it helpful to combine class and reading assignment notes addressing the same or related topics?

5.   As a valuable exercise in self testing, try to write a summary of the material in this reading assignment?

SUMMARY ▐███████████████████████

Self testing needs to be applied after each class and immediately after taking notes from reading assignments. Self testing also needs to be conducted after completion of certain segments of a course and periodically throughout the term when all notes taken to date are combined and reviewed. When reviewing notes, it is important to recite information aloud without looking at your notes in order to check your recall.

Strategy 10
Test Strategies

Test strategies enable students to maximize performance while preparing for and taking tests, thus increasing academic success. Application of test strategies enables students to determine in advance what to study and how to prepare for tests of different types as well as what steps to follow while actually taking tests.

APPLYING THIS STRATEGY

To apply this strategy, begin now to study for a test over this section of the book on learning strategies. What type of test do you think it will be? Can you predict some of the questions? With whom would you like to study? Reflect on these questions as you read the remainder of this section of the book.

PERFORMANCE OBJECTIVE

When you have completed Strategy 10, you will be able to specify the steps to take while preparing for tests and the recommended guidelines to follow while taking tests.

Test Strategies

The evaluation of a student's work depends greatly on his/her performance on tests. Tests are devised to check students' knowledge of the concepts and facts and their ability to critically analyze and evaluate the material presented. Some students consistently score better than their knowledge would seem to indicate, while other students usually score worse than they should, given their knowledge of a subject. The primary difference between these two types of student groups is their skill in taking tests.

Test strategies can be divided into two categories. The first category consists of those skills that are involved in preparing for a test. Test preparation involves application of all nine study strategies presented so far. Test preparation also involves a sequence of steps which includes determining in advance what will be covered and the type of test to be administered.

The second category of test strategies involves those steps to follow while taking a test. When taking a test, the student wishes to maximize her/his score. In general, the following guidelines apply:

Read test directions carefully and survey the entire test before beginning.
Budget time so all items can be completed.
Read questions carefully and answer those you're sure of first.
Save time to review answers before submitting the completed test.

Two lists of attributes are shown below. Each item illustrates some attribute of a student whose ability to apply test strategies helps or hinders his/her academic performance. A blank space appears before each example. If that example seems to fit your situation, place a check mark (✓) in the blank.

Likely Attributes of Effective Application of Test Strategies

____ Knows, if at all possible, what type of test will be given
____ Asks, on a multiple-choice test, whether wrong responses count against the final score
____ During the test, regularly makes note of time remaining and test items still to be completed
____ Is rarely surprised by a test question

Likely Attributes of Ineffective Application of Test Strategies

____ Fails to complete many tests in the time allowed
____ Lacks understanding of how to respond appropriately to essay items although he/she knows the information required for the response
____ Always answers every test item in sequential order
____ Finds some test questions that were expected, but finds several that he/she hoped wouldn't be asked

Review: Test Strategies

Let's review what you need to know in order to apply test strategies effectively.

1. In order to prepare adequately for an upcoming test, you should follow certain guidelines. Initially, you should discover, at least a week in advance if possible, what kind of test will be given and what kinds of items will be included. For example, if a test

will contain numerous multiple-choice, true-false, or fill-in-the-blank items, you might focus much of your study on terms and their definitions, names, dates, places and events, and concepts succinctly articulated. If numerous essay items will be included, you need to make sure you know the main points covered and all related major details in order to prepare responses adequately. You may need to explore relationships among the concepts covered, create concept maps, draw diagrams, and construct time lines in order to review the "whole picture." Further, you should be familiar with the format of essay responses and what is meant by direction words such as explain, compare and contrast, summarize, and the like.

2. Once you have determined what will be covered and the type of items to be included, you need to review and study the material with an eye toward the type of items expected.

3. For all tests, it is wise to combine your material into some type of outline/concept map and recite several times, without referring to your notes, the information included in order to check your recall. Reciting information aloud helps you understand and remember what you need to know.

4. When actually taking a test, you need to read directions carefully and survey the entire test before answering any items. Read the questions carefully and answer those you're sure of first. Throughout the test period, budget your time as you proceed so you will be able to complete all test items. If possible, it is recommended to review your responses before submitting your test.

EXERCISES

There is one exercise for Test Strategies. Respond to the items below and on the following page in the spaces provided. Write your responses in complete sentences.

EXERCISE 1

1. What are two basic types of test items?

2. Why is it important to discover in advance the kind of items to be included on a test?

3.   When you review material for a test, why is it helpful to recite information aloud while covering your notes?

4.   When taking a test, what are three guidelines which should be followed?

SUMMARY

Test strategies involve both preparing for and taking tests. Guidelines recommended for preparing for tests include discovering in advance what will be covered and what kind of items will be included. Objective and essay items may be included. These types of items place different demands on the student when preparing for a test. One recommendation for preparing for all tests is to combine material into some type of outline or concept map and recite the material aloud in order to check recall, clarify understanding, and facilitate recollection.

Certain guidelines are recommended for taking tests too, including reading directions and surveying the entire test at the start, reading all questions carefully, and responding to those you know first. Budget your time throughout the test period and review responses before submitting the completed test. In the second section of this book, Skill 8, Taking Tests, will provide you the opportunity to practice application of test strategies.

The
Sciences

1
Surveying Textbooks

Before you read or study your science textbook, you should survey it. Survey means to "look over" or "examine." You survey a textbook to see what the book is about and how it is organized.

APPLYING THIS SKILL

To practice applying this skill, you need:
1.   This book
2.   A pen or pencil

Take these items to a work area. Read the review section to make sure you understand the information you need in order to apply this skill. Then complete the exercises for Skill 1.

PERFORMANCE OBJECTIVE

When you have completed Skill 1, you will be able to survey science textbooks. You will be able to locate and identify the parts that are found in the front and back of most science textbooks. You will be able to use the table of contents, glossary, and index to locate information when you survey science textbooks and when you study.

Surveying Textbooks

Before you read or study your science textbook, you should survey or examine it. This will give you an idea of how it is organized and of the parts it contains. Surveying a textbook means to read the parts found in the front of the book: title page, copyright notice, table of contents, and preface. It also means to look in the back of the book to see if there are the following parts: an appendix section, a glossary, a bibliography, a suggested reading list, and an index.

Let's review the information you need in order to apply this skill.

Front Parts

1. The *title page* tells you the title of the book, the subtitle if there is one, giving further information about the title, and the author's (or authors') name. It also tells you the edition of the book, the publisher of the book, and the city where the book was published.

2. The *copyright* notice is usually on the back of the title page. The copyright notice tells you the date when the book was published. If more than one date is given, it means that the textbook has been revised. The most recent date is the date when your textbook was published. The other dates given are dates of earlier editions or printings.

3. The *table of contents* tells you the major divisions of the textbook and the pages on which they begin. It shows you how the body of the textbook is organized. It also gives you an idea of the material covered.

    Some science textbooks are divided into several parts, units, or sections about general topics. Each part consists of several chapters which are numbered as usual. In science textbooks, the subjects listed under the chapter may also be numbered. When this is the case, each subject often begins with the chapter number, followed by a decimal or dash and the number of the subject. For example, the second subject listed under Chapter 5 is numbered 5-2.

4. The *preface* is a message which introduces the book to you. It usually tells you for whom the book is written, the author's purpose for writing it, how the book is organized, and other general information about the book. The preface may also be called an "Introduction," a "Foreword," or "To the Student."

Back Parts

1. An *appendix* section may contain information that helps make the material in the body of the textbook clearer. It may also contain information not discussed at length in the textbook but of importance to most science students. The body of the textbook may direct you to an appendix by telling you to "See Appendix." If there is more than one appendix, each one is usually given a letter, number, or Roman numeral.

2. Science textbooks sometimes have a *glossary*. In the glossary are many of the important terms discussed in the textbook. These terms are listed in alphabetic order and are defined for you. The correct pronunciation of the terms may be provided.

3. Either a *bibliography* or a *suggested reading list* may be included in science textbooks. A bibliography is a list of books and articles that the author used in preparing the textbook. A suggested reading list directs you to additional books and articles about the subjects covered in the textbook. Both lists can be used to find more information about these subjects. In addition, both lists may appear at the end of the chapters rather than in the back of the textbooks. In science textbooks, these lists usually do appear at the conclusion of the chapters.

    When included, the bibliography may be called "References." The suggested reading list may be called "Selected Readings," "Supplementary Readings," or "Suggestions for Further Reading."

4. Almost all science textbooks contain an *index*. The index is a list of specific topics and names discussed in the body of the textbook. The topics and names are listed in alphabetic order. They are followed by the page number or numbers on which they're discussed in the textbook. If only one page number is given, the topic may only be mentioned on that page. Page numbers with a dash between them means that a topic is discussed in detail on those pages. Often one or more subtopics are listed in alphabetic order under the topics. Specific page numbers follow subtopics as well.

    If the word "See" followed by another topic is listed instead of page numbers, you should look for that topic in the index to find the pages where it is discussed. If you find the words "See also" followed by another topic(s), before or after page numbers given, you may look for those topics in the index to find more/additional pages that discuss the given topic.

EXERCISES ▮▮▮▮▮▮▮▮▮▮▮▮▮▮▮▮▮▮▮▮▮

There are 16 exercises for Skill 1. To complete the exercises, you will use pages from science textbooks to answer some questions. Write your answers in the spaces provided in your book. You can check your answers with the answer key at the back of this book.

EXERCISE 1 ▮▮▮▮▮▮▮▮▮▮▮▮▮▮▮▮▮▮▮▮

Use the title page at the right to answer Questions 1-4.

1.   What is the title of this textbook?

2.   Does this textbook have a subtitle?
     If yes, what is the subtitle?

3.   Who are the authors of this textbook?

4.   Who published this textbook?

UNDERSTANDING

BIOLOGY

Burton S. Guttman The Evergreen State College
Johns W. Hopkins III Washington University

HBJ
Harcourt Brace Hovanovich, Inc.

New York San Diego Chicago San Francisco
Atlanta London Sydney Toronto

Source: From Burton S. Guttman and Johns W. Hopkins III, **Understanding Biology** (Orlando: Harcourt Brace Jovanovich, Inc., 1983), page iii (title page). Reproduced by permission of the publisher.

EXERCISE 2 ████████████████████████████████████

Use the copyright notice below to answer Questions 5-6.

5.   When was this textbook published?

6.   Who owns the right to publish the material in this textbook?

Source: From Arthur Beiser, **Modern Technical Physics,** fourth edition
(Menlo Park, CA: The Benjamin/Cummings Publishing Company, Inc.,
1983), page ii (copyright notice). Reprinted by permission of the publisher.

EXERCISE 3 ██████████████████████████

Part of a textbook table of contents is shown at the right. Use it to answer Questions 7-11.

7. What is Part Two of this textbook about?

8. On what page does Chapter 6 begin?

9. How many chapters are in Part Two?

10. What is Part Three about?

11. What is the title of Chapter 10?

Source: From Thomas C. Emmel, **Worlds Within Worlds: An Introduc-tion to Biology** (Orlando: Harcourt Brace Jovanovich, Inc., 1977), page viii. Reproduced by permission of the publisher.

EXERCISE 4 ██████████████████████████████████

Part of a textbook preface is shown below. Use it to answer Questions 12-13.

12.   The preface tells you the purpose of this book. What is the purpose of this book?

13.   Is there a study guide to accompany this textbook?

PREFACE

This textbook is intended to provide a comprehensive introduction to the nature and functioning of living systems within the biosphere on earth. This book begins with an analysis of the entire biosphere as a total unit. It then investigates in finer and finer detail how the parts of this biosphere function. This approach has proved to be the most stimulating and relevant method of presenting biology. It has been used successfully with thousands of University of Florida students.

There are two important features of this book's organization. First, it allows students to explore the most absorbing and exciting subjects in biology. Second, it urges students then to pursue the details of how these phenomena function down to the interacting parts of the living organism.

While we feel this overall approach — going from the grand scale of biosphere biome to the biological organization of the individual — will be most effective with students, the text does offer the instructor considerable flexibility. Each major chapter grouping is self-contained with clear definitions and can be used in the instructor's preferred order.

There are also a number of useful aids provided. A complete glossary reviews definitions of all important terms introduced in the text. Color inserts clarify major subjects discussed. A Study Guide, which includes extensive review material and study questions for each of the 34 chapters in this text, is available for students. An Instructor's Manual contains useful suggestions for lecture organization, appropriate films, and testing approaches. A Laboratory Manual written especially for students using this textbook is also available.

Source: From Thomas C. Emmel, **Worlds Within Worlds: An Introduction to Biology** (Orlando: Harcourt Brace Jovanovich, Inc., 1977), pgs. v and vi. Reproduced by permission of the publisher.

EXERCISE 5 ▐███████████████████████████████

Part of a textbook glossary is shown below. Use it to answer Questions 14-15.

14.   What is the definition of aerobe in this glossary?

15.   What is the definition of alveolus in this glossary?

adrenal medulla The inner portion of the adrenal gland

adrenalin A hormone produced by the adrenal medulla that stimulates a number of boy reactions

aerobe A microorganism that requires oxygen for respiration

allele One of several alternative gene forms at a particular position (locus) on homologodus chromosomes

allopatric species Species that live in different geographic areas; see sympatric species

alveolus One of the air sacs in the lungs

amino acid One of the organic acids that are the building blocks of protein molecules

amphibian One of the class of vertebrates between fish and reptiles that must return to the water to lay their eggs.

EXERCISE 6 ▰▰▰▰▰▰▰▰▰▰▰▰▰▰▰▰▰▰▰▰▰▰▰▰▰

Part of a textbook index is shown at the right. Use it to answer Questions 16-20.

16. Find the topic "Radicle."
     On what pages will you find
     information about radicles?

17. Find the topic "Pulse-chase
     experiments." Which pages
     discuss pulse-chase experiments
     in most detail?

18. Find the topic "Pyruvate."
     Under what topic will you find
     the page numbers where pyruvate
     is discussed?

19. Find the topic "Regulation."
     Under what other topic will you
     find more pages that discuss regulation?

20. Find the topic "Reactions."
     How any subtopics are listed
     under reactions?

Ptolemy, 10-11
Puberty, 257
Puccinia graminis (wheat rust), cycle
    of, 262
Pulse-chase experiments,
    292-293, 295
Punctuated equilibrium, 48-49
Pyrophosphate, 171
    importance in biosynthesis, 211
Pyrrole ring, 183
Pyruvate, *see* Pyruvic acid
R
Radiation
    biological effects of, 299-300
    effects of, 228
    infrared: perception of, *602*
    and temperature in plants, 459-461
Radicle, 444-445
Radioisotopes, 150
    in geological dating, 26
Reaction cascades, 557-558
    in neurons, 633
Reaction center, din photosynthesis,
    200-201
Reactions
    chemical, 123-125, 141-144
    coupled, 127-128
    cyclic, 180
Receptive field, 620-621, 622-623
Regeneration
    capacity for, 404
    in *Planaria*, *862*
Regulation; *see also* Homeostasis
    of enzyme activity, 156
    genetic, 160
    of major pathways, 190
    through different coenzymes, 186
Regulator gene, 352
    transposable type, in maize, 417
Regulatory mechanisms, 152

Source: From Burton S. Guttman and Johns W.
Hopkins III, **Understanding Biology** (Orlando:
Harcourt Brace Jovanovich, Inc., 1983), page 970.
Reproduced by permission of the publisher.

EXERCISE 7 ▐▬▬▬▬▬▬▬▬▬

Use the title page at the right to answer Questions 21-25.

21. What is the title of this textbook?

22. Does this textbook have a subtitle?
     If yes, what is the subtitle?

23   Who is the author of this textbook?

24. What edition is this textbook?

25. Who published this textbook?

FOUNDATIONS OF COLLEGE CHEMISTRY

MORRIS HEIN   5th EDITION

MOUNT SAN ANTONIO COLLEGE

Brooks/Cole Publishing Company
MONTEREY, CALIFORNIA

Source: From Morris Hein, **Foundations of College Chemistry**, fifth
edition (Monterey, CA: Brooks/Cole Publishing Company, 1982), page i
(title page). Reprinted by permission of the publisher.

EXERCISE 8 ▐███████████████████

Use the copyright notice below to answer Questions 26-27.

26.   When was the most recent edition of this textbook published?

27.   Who owns the right to publish the material in this textbook?

Source: From Stephen H. Spurr and Burton V. Barnes, **Forest Ecology**,
third edition (New York: John Wiley and Sons, 1980), page iv (copyright
notice). Reprinted by permission of the publisher.

EXERCISE 9

Part of a textbook table of contents is shown at the right. Use it to answer Questions 28-32.

28. What is Chapter 30 of this
     textbook about?

29. On what page does
     Chapter 3l begin?

30. How many sections are in
     Chapter 32?

31. What is Appendix A about?

32. What is the title of
     Appendix B?

Source: From Arthur Beiser, **Physics**, third edition (Menlo Park, CA: The Benjamin/Cummings Publishing Company, Inc., 1982). Reprinted by permission of the publisher.

EXERCISE 10 ▬▬▬▬▬▬▬▬▬▬▬▬▬▬▬▬▬▬▬▬▬▬▬▬▬▬▬

Part of a textbook preface is shown below. Use it to answer Questions 33-34.

33.   What background do you need to understand this textbook?

34.   The preface tells you the purpose of this book. What is the purpose of this book?

PREFACE

This textbook is directed at students who will need some compe-tence in physics in their careers. It is written to provide them with a basic foundation in physics. The organization is straight-forward, from the laws of motion to the properties of matter in bulk, optics, and finally modern physics. Mathematics have been kept to a minimum. Only elementary algebra and simple trigonometry are used. Students will need to have background in those areas of math to understand this text. Math areas are reviewed, however, in an appendix; so are powers-of-ten notation.

About 600 illustrations are provided to help make the reader's task easier. Marginal notes, sample problems, and lists of important terms and formulas are provided at the end of each chapter too. The more than 2000 exercises are divided into three categories: mul-tiple-choice questions; exercises for practice in manipulating numbers and formulas; and problems. Answers to all multiple-choice questions are provided in the text. Outline solutions for the odd-numbered exercises and problems are given in the text also.

Outline solutions for even-numbered exercises and problems are given in the Solutions Manual. Each outline solution shows how the problem can be attacked and gives the answer, but most of the intermediate steps are omitted. In this way the reader can be guided, but still have something left to do on his or her own.

New in the Second Edition

The emphasis of the book on the fundamentals of physics remains unaltered. A number of changes, however, were made to better serve students in a variety of disciplines. In particular, the applications of physics in various aspects of the life sciences now receive more attention. Such topics as muscular forces, blood pressure and animal metabolism are discussed in this edition. New material is provided in other areas as well. For example, fluid flow and blackbody radiation are given expanded treatment. To make room for these additions, a number of subjects of lesser interest were shortened.

Source: From Arthur Beiser, **Physics**, second edition (Menlo Park, CA: The Benjamin/Cummings Publishing Company, Inc., 1978), preface. Reprinted by permission of the publisher.

EXERCISE 11

Part of a textbook glossary is shown below. Use it to answer Questions 35-36.

35.   What is the definition of Covalent Bond in this glossary?

36.   What is the definition of Doppler Effect in this glossary?

Covalent Bond. In a covalent bond between adjacent atoms of a molecule or solid, the atoms share one or more electron pairs.

Crystalline Solid. Solids whose constituent atoms or molecules are arranged in regular, repeated patterns are called crystalline. When only short-range order is present, the solid is amorphous.

Dispersion. Dispersion refers to the splitting up of a beam of light containing different frequencies by passage through a substance whose index of refraction varies with frequency.

Domain. An assembly of atoms in a ferromagnetic material whose atomic magnetic moments are aligned is called a domain.

Doppler Effect. The Doppler effect refers to the change in frequency of a wave when there is relative motion between its source and an observer.

Effective Value. The *effective value* of an alternating current is such that a direct current of this magnitude produces heat in a resistor at the same rate as the alternating current. The relationship between i_{eff} and i_{max} is $i_{eff} = 0.707 i_{max}$.

Source: From Arthur Beiser, **Physics**, second edition (Menlo Park, CA: The Benjamin/Cummings Publishing Company, Inc., 1978), page 841. Reprinted by permission of the publisher.

EXERCISE 12 ████████████████████████████████

Part of a textbook index is shown at right. Use it to answer Questions 37-41.

37. Find the topic "quarks."
     On what page/s will you find
     information about quarks?

38. Find the topic "Pollution."
     Which pages discuss pollution
     in most detail?

39. Find the topic "Positive charge."
     Under what topic will you find
     the page numbers where positive
     charge is discussed?

40. Find the topic "Power." Under
     what other topic will you find
     more pages that discuss power?

41. Find the topic "Quantum theory."
     How many subtopics are listed
     under Quantum theory?

Planck's quanium hypothesis, 422-24
Plato, 20
Pluto, discovery of, 79
Polarization of light, 328-30
Pollution, 245, 403-l0, 483, 487
Positive charge *See* Electric charge
Positron, 467, 489
Pound (unit), 54
Power, l23-24, 36l-63. *See also* Nuclear power
Pressure, l76-83, l85-93
     atmospheric, l85-89
Primary colors, 335-36
Probability distributions, 440
Projectile motion, 66–70
Proton, 462
Ptolemy, 2l-22
Pythagoras, l9

Quantum of light. *See* Photon
Quantum mechanics, 438-48
     philosophy of, 442-46
Quantum number, 435
Quantum theory, 422-55
     of atomic structure, 43l-38, 446-47
     of light, 425-27
     of molecular structure, 447-48
Quarks, 490-9l

Source: From Douglas C. Giancoli, **The Ideas of
Physics,** second edition (Orlando: Harcourt Brace
Jovanovich, Inc., l978), page 526. Reproduced by
permission of the publisher.

EXERCISE 13 █████████████████████████████

Use the title page at the right to answer Questions 42-45.

42.   What is the title of this textbook?

43.   Does this textbook have a subtitle?
      If yes, what is the subtitle?

44.   Who is the author of this textbook?

45.   Who published this textbook?

PHYSICS:

For Scientists and Engineers

Raymond A. Serway
James Madison University

*Cover photo by
Ross Chapple, Hume, Virginia*

SAUNDERS GOLDEN SUNBURST SERIES

SAUNDERS COLLEGE PUBLISHING
Philadelphia New York Chicago
San Francisco Montreal Toronto
London Sydney Tokyo Mexico City
Rio de Janeiro Madrid

Source: From: **Physics: For Scientists and Engineers,** by Raymond A. Serway. Copyright (c) 1982 by Raymond A. Serway, page ii (title page). Reprinted by permission of CBS College Publishing.

EXERCISE 14 ████████████████████████████████████

Part of a textbook table of contents is shown at the right. Use it to answer Questions 46-50.

46. What is Chapter 1 of this textbook about?

47. On what page does subsection 1-3 begin?

48. How many major sections are in Chapter 1?

49. What is Chapter 2 about?

50. What is the title of subsection 2-2?

Contents

Source: From Raymond E. Davis, Kenneth D. Gailey, and Kenneth W. Whitten, **Principles of Chemistry** (Philadelphia: W. B. Saunders College Publishing, 1984), page xvii. Reprinted by permission of the publisher.

EXERCISE 15 ▮▬▬▬▬▬▬▬▬▬▬▬▬▬▬

Part of a textbook index is shown at the right. Use it to answer Questions 51-54.

51. Find the topic "Permanent waving." On what page/s will you find information about permanent waving?

52. Find the topic "Polysaccharides." Which pages discuss polysaccharides in most detail?

53. Find the topic "Sleeping pills." Under what topic will you find the page numbers where sleeping pills are discussed?

54. Find the topic "Safety with chemicals." How many subtopics are listed under Safety with chemicals?

Pencillin, 396-399
Periodic Table, 61, 62-64, 89-92,
    105, 161
Permanent waving, 378-380
Perspiration, 377
Polymers, **190** (*see also* Addtion
    polymers, Condensation
    polymers
Polysaccharides, 218, 222-227
Powders:
    cosmetic, 375
    gun, 155, 156, 189, 314
Proton, 67, 71-73
    charge of, 73
    weight of, 73
Safety with chemicals:
    aerosols, 30-371
    agricultural chemicals, 302
    childpoisoning, 384-385
    detergents, 361-362
    drain cleaners, 366-367
    lead in paints, 372
    mixing cleaners, 367-368
    oven cleaners, 368
    petroleum solvents, 367
    toilet bowl cleaners, 366
Salicylamide, 390
Silicone, 195
Silver, sterling, 148
Sleeping pills (*see* Barbiturates)
Sludge, sewage, 451-453
Smelting, 319,323

Source: From Tom Hughes, **Chemistry: Ideas to Interpret Your Changing Environment** (Encino, CA: Dickenson Publishing Company, Inc., 1975), page 499. Reprinted by permission of the publisher.

EXERCISE 16 ████████████████

Part of a textbook glossary is shown below. Use it to answer Questions 55-56.

55. What is the definition of gamete in this glossary?

56. What is the definition of genetic map in this glossary?

G

gametangium In organisms such as fungi and algae, the structure in which gametes are formed.

gamete One of the reproductive cells of a sexually reproducing organism.

gastric Referring to the stomach.

gene A unit of inheritance, generally defined now as a portion of a nucleic acid that informs a single polypeptide chain.

genetic information The sum total of the information carried in the genome of an organism.

genetic map A representation of the structure of a genome as determined by mapping experiments.

genome The structure of an organism that contains instructions for the organism's growth and reproduction.

genotype A description of the particular set of genes that an organism carries, as contrasted with its appearance (see phenotype).

Source: From Burton S. Guttman and Johns W. Hopkins III, **Understanding Biology** (Orlando: Harcourt Brace Jovanovich, Inc., 1983), page 931. Reproduced by permission of the publisher.

SUMMARY

1. When you get a textbook, take a few moments to survey or examine it. You will see what the book is about and how it is organized. When you survey a textbook, locate and examine the title page, copyright notice, and table of contents in the front of the book. Read the preface. Look in the back of the book for the appendix, glossary, bibliography or suggested readings list, and index. Check to see what features these parts include.

2. The title page is at the front of a textbook. You read the title page to make sure you know the exact title and to learn details about the title, author, and publisher. The title page will also tell you the edition of the book.

3. The copyright notice is also in the front of a textbook, usually on the back of the title page. You look at the copyright notice to find out the year or years the book was published. The copyright notice also tells you who owns the right to publish the material in the book and, often, whom to write for permission to copy material.

4. The table of contents is also found in the front of a textbook. It lists the major divisions of the book and the pages on which each chapter and section begins. You look at the table of contents to get an idea of the material covered in the book and how it is organized.

5. You may also find a preface, or introductory message, in the front of a textbook. A preface may be called "Introduction," "Foreword," or "To the Student." You read the preface to find general information about the author's purpose in writing the book, for whom the book is written, and how it is organized.

6. Many textbooks contain an appendix in the back of the book. The appendix gives additional information about topics covered in the body of the book or reference material that you can use as you are studying the chapters. The body of the textbook usually has an instruction that refers you to the appendix. If there is more than one appendix, each appendix is usually given a number, a letter, or a Roman numeral.

7. Some textbooks include a glossary in the back. A glossary is a list of special terms used in the body of the book. The terms are listed in alphabetic order with their definitions or meanings. Sometimes the glossary also gives the pronunciations of these terms and the pages on which the terms are first used.

8. You may also find a bibliography or a suggested reading list. These lists may appear either in the back of a textbook or at the end of each chapter or part in the body of the textbook. A bibliography is a list of books and articles that the author used in writing the textbook. The books and articles in either a bibliography or a suggested reading list will give you more information about the subjects in the textbook.

9. If a textbook contains an index, it is found in the back of the book. An index is an alphabetic list of specific topics covered in the body of the textbook. The index tells you the pages on which each topic listed is discussed. Sometimes the words "See" and "See also" tell you which topic(s) to check for page numbers. Using the index is the fastest way to find specific topics in the body of the textbook.

Additional Suggestions for Surveying Science Textbooks

Study guides and laboratory manuals may accompany science textbooks. They are designed to help you master the concepts and terms presented in the textbook.

In a study guide, there are usually the same number of chapters as there are in the textbook, one for each chapter of the book. Each chapter in the guide contains a variety of exercises, for example, multiple choice, true-false, and essay questions. These exercises may require you to know definitions and facts. They may also require you to apply scientific concepts and to write equations. Some problems involving mathematical computation may be included too. Formulas may need to be applied to solve some problems.

Some self-study materials are available for certain courses too, such as introductory physics courses. These materials may not be related to specific textbooks but treat the same topics covered in the books designed for such courses.

A laboratory manual is designed for use in the lab associated with a course. The lab manual contains information and directions necessary for completing related experiments. For each experiment a summary of the topic and any related equations are provided; a description of the equipment to be used and how to use it follows; and finally, space is allotted for students to write their observations and conclusions. For science courses, however, instructors may prepare the directions and format for noting experimental results rather than use lab manuals published with textbooks. In the event that the textbook is changed, the same lab set-up directions, equipment, and material can still be used.

Make sure you survey the study guide, laboratory manual, or any self-study material that may be available when you survey the textbook for the course.

2
Surveying Chapters

When a textbook chapter is assigned for you to read, the first thing you should do is spend a few minutes surveying it. By surveying a chapter before you read it, you see what the chapter is about and how it is organized. You also learn what the author thinks is the most important information in the chapter. Using the information gained from surveying, you can set a clear purpose for reading the chapter.

APPLYING THIS SKILL

To practice applying this skill, you need:
1. This book
2. A pen or pencil
3. Blank paper

Take these items to a work area. Read the review section to make sure you understand the information you need in order to apply this skill. Then complete the exercises for Skill 2.

PERFORMANCE OBJECTIVE

When you have completed Skill 2, you will be able to survey chapters in science textbooks. You will be able to specify what the chapter is about through reading chapter openings, headings, and closings. You will be able to show how the chapter is organized by outlining the headings in a chapter. You will be able to specify a purpose for reading the chapter.

Surveying Textbook Chapters ▮▮▮▮▮▮▮▮

When a chapter is assigned for you to read, the first thing you should do is spend a few minutes surveying it. To survey a chapter, you

1.   Read the chapter opening.
2.   Examine the body of the chapter.
3.   Read the chapter closing.

Now let's review what you need to know in order to apply this skill.

1.   The chapter opening includes the title of the chapter which tells you what the chapter is about. It also includes one or more of the following: questions, objectives, introductory paragraph(s). It may include an outline of the chapter's content. These parts tell you more specifically what the chapter is about.

2.   The body of the chapter, which follows the opening, is usually divided into sections. Each section has its own heading which tells you what the section is about. Headings are easy to see because of their special size, color, darkness, or position. Each section with a heading is usually divided into subsections with smaller headings.

3.   When you examine the body of the chapter, you read all of the headings. They give you more information about what's discussed in the chapter and show how the chapter is organized. When you examine the body of the chapter, use the headings as clues to how the sections are related. To show how all the sections are related, you can write the headings in outline form. In a simple outline you write the number and title of the chapter at the top of the page. Then you put the heading of each main section next to the left margin. You indent the heading(s) of each section's subsection(s) under it. Following is an outline of chapter headings from a biology textbook:

Chapter 10: Respiration	(Chapter title)
Respiration in Water	(Main section heading)
Gas exchange: simple animals	(Subsection headings)
Gas exchange: animals with gills	"
Respiration on Land	(Main section heading)
Diffusion through skin	(Subsection headings)
Tracheal system	"
Lung system	"

A subsection may be further divided into subsections with even smaller headings. To place these headings in the outline, you would indent them under the subsection heading.

4. After examining the body of the chapter, you read the chapter closing when surveying a chapter. The chapter closing sums up or reviews the main points covered in the chapter. The chapter closing often consists of one or more summary paragraphs and a list of review questions or exercises. The various parts of the chapter closing tell you what ideas, terms, or facts are considered important by the author. Reading the closing helps you see what information you should pay attention to when you read the chapter. Doing so helps you establish a clear purpose for reading the chapter thoroughly.

In science textbooks, the closing often contains questions requiring you to apply scientific principles. Problems requiring the use of formulas and mathematical computation may be included too, particularly in physics and chemistry. Often solutions to the odd-numbered problems are provided in the back of the textbook.

5. In setting a purpose for reading a chapter, you use the information which you gain through surveying, essentially the chapter's main topics or points. To establish your purpose, you specify what you want to learn about those topics or points. Sometimes it may be important to learn how something is done or how something happens, what evidence or important facts support specific theories, why certain conclusions have been reached, or how to use formulas and mathematical computations to solve problems. Sometimes you may need to learn the meanings of certain terms and how

they are used or applied. Specifying in advance why you will
read the chapter helps you to increase your understanding of
the material when you read it, and it helps you to remember
the material later.

Below is an outline of the headings from another chapter in a biology textbook.

Chapter l3: Evolution

Evidence for Evolution
> Fossil Remains
> Ontogeny
> Subspeciation

Three Theories of Evolution
> Natural Selection
>> Macroevolution
>> Microevolution
> The Neo-Darwinian Theory of evolution
> Saltational Evolution
>> Species-Specific Gene Complexes
>> Polymorphism

The Process of Adaptation
> Formation of Races
> Adaptive Radiation
> Adaptive Polymorphism

Isolating Mechanisms

EXERCISES

There are 9 exercises for Skill 2. In each exercise, you will survey a brief chapter in the science section of
this book. Each sample chapter treats topics typically covered in science textbooks. You will answer
questions about the chapter opening and about the chapter closing. You will show how the chapter is
organized by outlining the chapter's headings. You will use the information gained through surveying the
chapter to specify a purpose for reading the chapter.

For Exercises l-3, you will use chapters treating topics from biology textbooks. For exercises 4-6, the
chapters are drawn from physics textbooks. For Exercises 7-9, you will work with chapters treating topics
from chemistry textbooks. You may complete as many exercises as you need, and you may work on the
exercises in any order.

The directions at the start of each exercise tell you which sample chapter to use. Be sure to find the correct
chapter in this book before you begin answering the questions. Write your answers to the questions in the
spaces provided in this book; use your own paper to write an outline of the chapter headings. You can
check your answers with the answer key at the back of this book.

EXERCISE 1 ▮▮▮▮▮▮▮▮▮▮▮▮▮▮▮▮▮▮▮▮▮▮▮▮▮

The chapter for this exercise is on pages 215-217.

1.   What is the title of Chapter 12?

2.   In addition to the chapter title, which of these parts is/are included
     in the opening of Chapter 12? Circle the letter or letters of the answer.
     a. questions
     b. objectives
     c. introductory paragraph(s)

3.   Based on the chapter opening, what is the chapter about? Write your answer in one
     or two sentences.

4.   Show how the body of the chapter is organized by writing the headings in outline form.
     Use your own paper.

5.   According to the summary, what gives blue-green algae their color?

6.   According to the summary, where are blue-green algae commonly found?

7.   According to the review questions, which of these topics is/are important in this chapter?
     Circle the letter or letters of the answer.
     a. how blue-green algae reproduce
     b. what chemicals are used to destroy blue-green algae
     c. where blue-green algae are found
     d. how blue-green algae produce carbon dioxide

8.   Based on surveying, what would be the major purpose(s) for reading this chapter?
     Write your answer in one or two sentences.

EXERCISE 2

The chapter for this exercise is on pages 219-222.

9.   What is the title of Chapter 13?

10.   In addition to the title, which of these parts is/are included
      in the opening of Chapter 13? Circle the letter or letters of the answer.
      a. questions
      b. objectives
      c. introduction

11.   Based on the chapter opening, what is the chapter about?
      Write your answer in one or two sentences.

12.   Show how the body of the chapter is organized by writing the headings in outline form.
      Use your own paper.

13.   Based on the summary, what are protozoa?

14.   According to the summary, ciliates have hair-like cilia on their surface.
      What are cilia used for?

15.   According to the review questions, which of these topics is/are
      in this chapter? Circle the letter or letters of the answer.
      a. how some amoebae can be dangerous to man
      b. how amoebae move
      c. how cilia are used
      d. how protozoa are classified

16.   Based on surveying, what would be the major purpose(s) for reading this chapter?
      Write your answer in one or two sentences.

EXERCISE 3 ██████████████████████████████

The chapter for this exercise is on pages 223-225.

17. What is the title of Chapter 11?

18. In addition to the title, which of these parts is/are included
    in the opening of Chapter 11? Circle the letter or letters of the answer.
    a. questions
    b. objectives
    c. introductory paragraph(s)

19. Based on the chapter opening, what is the chapter about?
    Write your answer in one or two sentences.

20. Show how the body of the chapter is organized by writing the headings in outline form.
    Use your own paper.

21. Based on the summary, what two types of pea plants did Mendel use in his first experiment?

22. According to the summary, Mendel's first experiment showed that all of the offspring
    (F_1 generation) had only one form of the trait. Based on this result, what was Mendel's
    conclusion?

23. According to the review questions, which of these topics is/are important in this
    chapter? Circle the letter or letters of the answer.
    a. breeding
    b. law of dominance
    c. male gametes
    d. dominant traits

24. Based on surveying, what would be the major purpose(s) for reading this chapter?
    Write your answer in one or two sentences.

EXERCISE 4 ███████████████████████████

The chapter for this exercise is on pages 227-230.

25.   What is the title of Chapter 2?

26.   In addition to the chapter title, which of these parts is/are included
       in the opening of Chapter 2? Circle the letter or letters of the answer.
       a. questions
       b. objectives
       c. introductory paragraph(s)

27.   Based on the chapter opening, what is the chapter about?
       Write your answer in one or two sentences.

28.   Show how the body of the chapter is organized by writing the headings in outline form.
       Use your own paper.

29.   Based on the summary, what is the speed of a moving body?

30.   According to the summary, what is the velocity of a body?

31.   Based on the exercises, which of these topics is/are important in this chapter?
       Circle the letter or letters of the answer.
       a. figuring amount of work done when the force and weight of an object are given
       b. figuring distance when mph and time are given
       c. figuring speed in mph when distance and time are given
       d. distinguishing between distance and speed

32.   Based on surveying, what would be the major purpose(s) for reading this chapter?
       Write your answer in one or two sentences.

EXERCISE 5 ▐███████████████████████████

The chapter for this exercise is on pages 231-235.

33. What is the title of Chapter 4?

34. In addition to the title, which of these parts is/are included
     in the opening of Chapter 4? Circle the letter or letters of the answer.
     a. questions
     b. objectives
     c. introductory paragraph(s)

35. Based on the chapter opening, what is the chapter about?
     Write your answer in one or two sentences.

36. Show how the body of the chapter is organized by writing the headings in outline form.
     Use your own paper.

37. Based on the summary, what does Newton's first law of motion tell us?

38. According to the summary, what does the third law of motion tell us?

39. Based on the exercises, which of these topics is/are important in this chapter?
     Circle the letter or letters of the answer.
     a. how Newton discovered the three laws of motion
     b. how one can increase force
     c. how Newton's second law of motion can be applied to the real world
     d. the definition of a newton (N)

40. Based on surveying, what would be the major purpose(s) for reading this chapter?
     Write your answer in one or two sentences.

EXERCISE 6 ████████████████████████████████

The chapter for this exercise is on pages 237-242.

41.   What is the title of Chapter 7?

42.   In addition to the title, which of these parts is/are included in the opening of
      Chapter 7? Circle the letter or letters of the answer.
      a.   questions
      b.   objectives
      c.   introductory paragraph(s)

43.   Based on the chapter opening, what is the chapter about?
      Write your answer in one or two sentences.

44.   Show how the body of the chapter is organized by writing the headings in outline form.
      Use your own paper.

45.   Based on the summary, what is the equation for work?

46.   According to the summary, what is the unit of work in the SI system?

47.   Based on the exercises, which of these topics is/are important in this chapter?
      Circle the letter or letters of the answer.
      a.   how to calculate the number of joules of work done
      b.   how to save energy
      c.   how to increase energy
      d.   how to distinguish positive work from negative work

48.   Based on surveying, what would be the major purpose(s) for reading this chapter?
      Write your answer in one or two sentences.

EXERCISE 7 ████████████████████████████████

The chapter for this exercise is on pages 243-245.

49.   What is the title of Chapter 3?

50.   In addition to the title, which of these parts is/are included
      in the opening of Chapter 3? Circle the letter or letters of the answer.
      a. questions
      b. objectives
      c. introductory paragraph(s)

51.   Based on the chapter opening, what is the chapter about?
      Write your answer in one or two sentences.

52.   Show how the body of the chapter is organized by writing the headings in outline form.
      Use your own paper.

53.   Based on the summary, what does the Law of Conservation of Mass tell us?

54.   According to the summary, in terms of atoms, of what does a molecule of an element consist?

55.   Based on the exercises, which of these topics is/are important in this chapter?
      Circle the letter or letters of the answer.
      a. atom
      b. particle
      c. molecule
      d. carbon

56.   Based on surveying, what would be the major purpose(s) for reading this chapter?
      Write your answer in one or two sentences.

EXERCISE 8 ■■■■■■■■■■■■■■■■■■■■■■■■■■

The chapter for this exercise is on pages 247-250.

57. What is the title of Chapter 5?

58. In addition to the title, which of these parts is/are included
     in the opening of Chapter 5? Circle the letter or letters of the answer.
     a. questions
     b. objectives
     c. introductory paragraph(s)

59. Based on the chapter opening, what is the chapter about?
     Write your answer in one or two sentences.

60. Show how the body of the chapter is organized by writing the headings in outline form.
     Use your own paper.

61. Based on the summary, what is the periodic table?

62. According to the summary, what are the vertical columns in the table?

63. Based on the questions, which of these topics is/are important in this chapter?
     Circle the letter or letters of the answer.
     a. Bohr's explanatiom of the existence of atomic spectra and the energy radiated
        by atoms by picturing the atom as a tiny positive nucleus with electrons in motion
        around it
     b. position of elements within the table
     c. discovery of specific elements
     d. rows and columns in the periodic table

64. Based on surveying, what would be the major purpose(s) for reading this chapter?
     Write your answer in one or two sentences.

EXERCISE 9 ▮▮▮▮▮▮▮▮▮▮▮▮▮▮▮▮▮▮▮▮▮▮▮

The chapter for this exercise is on pages 251-255.

65. What is the title of Chapter 7?

66. In addition to the title, which of these parts is/are included in the chapter opening?
    Circle the letter or letters of the answer.
    a. questions
    b. objectives
    c. introductory paragraph(s)

67. Based on the chapter opening, what is the chapter about?
    Write your answer in one or two sentences.

68. Show how the body of the chapter is organized by writing the headings in outline form.
    Use your own paper.

69. Based on the summary, what happens when pressure is increased?

70. According to the summary, what happens when pressure is decreased?

71. Based on the problems, which of these topics is/are important in this chapter?
    Circle the letter or letters of the answer.
    a. how to calculate the volume of gas when the pressure is increased
    b. how to calculate the volume of water (in a container) when the temperature is changed
    c. how to calculate the volume of gas when the pressure is decreased
    d. how to calculate the volume of gas when the temperature is decreased

72. Based on surveying, what would be the major purpose(s) for reading this chapter?
    Write your answer in one or two sentences.

SUMMARY

1. Surveying a chapter before you read it helps you see what the chapter is about and how it is organized. It also helps you focus on the important information when you read the material, and it helps you set a purpose for reading the entire chapter.

2. To survey a chapter, first read the chapter opening. Next, examine the headings in the body of the chapter. Then read the chapter closing.

3. The chapter opening begins with the number and title of the chapter. It may include introductory paragraphs, questions, and/or objectives.

4. The body of the textbook chapter is usually divided into sections. Each section has a heading which tells you what the section covers. Examine the headings to learn what information is covered in the chapter and how it is organized. The size, color, and position of the headings help show how the body of the chapter is organized.

5. To show how a chapter is organized, you can make an outline of the chapter's headings. In a simple outline, the chapter number and title are placed at the top in the center of the page. The main section headings are placed at the left margin. Under each main section heading, the subsection headings are indented from the margin. If a subsection is divided into smaller sections, the headings for these subsections are indented further from the left.

6. The chapter closing often includes a summary of the material covered in the chapter. The summary is a specific, detailed review of the most important points in the chapter. Most chapter closings also include a review section – usually a list of questions for discussion or review. In some textbooks, the chapter closing also includes key terms or exercises. By reading the chapter closing before you read the chapter, you will know what information to focus on when you study.

7. To set a purpose for reading the chapter, you use the information gained from surveying the chapter's main topics and points. You specify what you want to learn about those topics and points. You may need to learn how something happened, how certain formulas are applied, what evidence supports specific theories, why certain conclusions have been reached or the definitions of specific terms and how they're applied. Clarifying why you will read the chapter helps you increase your understanding of the material when you do read it and your later recall of that chapter's content.

3
Building Vocabulary

In order to understand clearly the textbook chapters assigned, you will need to know the meaning of the words, both technical terms as they are used in the sciences and other words used in context. With knowledge of the words used in the material, you can gain an accurate understanding of the concepts conveyed.

APPLYING THIS SKILL

To practice applying this skill, you need:
1. This book
2. A pen or pencil

PERFORMANCE OBJECTIVE

When you have completed Skill 3, you will be able to build vocabulary enabling you to understand the meaning of the words found in science textbooks. You will be able to determine meanings of words by applying context clues. You will be able to unlock meanings of words using word analysis and applying meanings of commonly used prefixes and roots. Also, you will be able to apply a systematic method of studying words (to be further developed when completing Skills 4 and 6 in this book).

Building Vocabulary ▬▬▬▬▬▬▬▬

When reading chapters assigned and listening to lectures when taking courses in the science areas, it is important to understand and use correctly the terms encountered. One of the best ways to increase your vocabulary in the sciences is to read much material addressing science topics and issues. In addition to science textbooks, several journals are available on the market and in the library; related books address similar topics also. When reading a great amount of material, you will meet some of the same terms in a science context and will come to know them. You will realize also that some of these terms have very different meanings in other contexts. Further, you will begin to expand your general vocabulary or knowledge of words used in a range of written material and dialogues.

In addition to reading a wide range of materials in the field, there are some direct steps you can take which will help you not only increase your science vocabulary but knowledge of word meanings in general.

Let's review what you need to know in order to apply this skill to words you meet when reading science material.

1. Several context clues are often incorporated in written material. These clues point to the meanings of specific words. You need to be sensitive to these clues in order to understand the meanings of words new to you. Below are some types of context clues used commonly in textbook material.

   a) Definition. Sometimes writers define words or use synonyms next to words, especially when introducing new terms crucial to the meaning of the concepts presented. An example of a definition provided in context is as follows:

   Photosynthesis is the process by which light energy is absorbed and then converted to the chemical bond energy of glucose.

   b) Example. Writers sometimes explain the meanings of terms by providing simple examples of the terms introduced. See below:

   Many microorganisms called anaerobes which do not use oxygen in respirations undergo **anaerobic respiration** or respiration in absence of air. For example, respiration in yeast is anaerobic respiration and is commonly called fermentation.

c) Contrast.   Writers may explain a term by contrasting it with
                            another term, opposite in meaning. See below:

The animal which did not have a specific direction of locomotion was not
said to have **bilateral symmetry**; when a cut down the midline from head
to tail was made, the halves were quite different.

d) Additional Information.   Sometimes writers provide further
                               details to clarify the meaning of a term.
                               See below:

The series of changes in a community during its development can be called
ecological succession. Succession may occur as a result of natural, orderly
changes, or it may follow a disaster such as fire or disease.

You need to be aware of these clues as you study a word in context, the
sentence, phrase and/or paragraph in which it appears. Words have meaning
as parts of sentences or paragraphs. The sentence itself has meaning in the
larger context of the paragraph. Further, terms have different meanings depending on the context. For example, the term blade in a given context may
mean the flat edged cutting part of a sharpened tool or weapon. It may refer
to a flat, thin part of a section: the blade of an oar. Then, it may refer to the
metal part of an ice skate. In anatomy, the balde is the scapula. In biology,
the blade is the expanded part of a leaf as distinguished from the leaf stalk.

As a student taking science courses, you need to be aware of differences
discussed above. You will need to learn definitely the meaning of such terms
as used in the sciences. Applying context clues will assist you with this task.

2. Another way to unlock meanings is to use word analysis by breaking a word into its parts; many words contain a prefix, root, and suffix. In order for this effort to be useful, you must know the meaning of these word parts, especially of the prefixes and roots. The root is the main part of a word. It provides the essence of the word's meaning. Prefixes are placed before the main part of a word or at the beginning of the word. For example, the root poper (Latin), meaning to place or put, forms the base of our word position, a place or location. When the prefix con, meaning with or together, is placed in front of position, the word created is composition, which refers to a putting together of parts or elements to form a whole; in science, composition may refer to the result or product of composing; mixture; compound. Further, when another prefix de, meaning down/away, is placed in front of composition, the word then formed is decomposition. In science, decomposition refers to the breakdown of materials of once living things and their wastes by bacteria or other tiny living things into simpler forms so they can be used again.

Some common prefixes which you should know are the following:

Prefix	Meaning	Example
Pre	Before	Precaution
Pro	Forward	Progress
Un	Not	Unhappy
In/Im/Il/Ir	Not	Independent
		Immune
Note: Im (before m, b, p)		Illusion
Il (before l)		Irreversible
Ir (before r)		
Non	Not	Nonconformist
Dis	Negation/Reversal	Disallow
In/Im/Il/Ir	In	Innate
		Immerse
		Illuminate
		Irrigate
Ex	Out	Exit
Re	Back/	Reply
	Again	Reassure
De	Down/	Descend
	Away	Departure
Sub	Under	Substandard
Trans	Across	Transaction
Mis	Wrong	Misinterpret
Inter	Between	Interferon
Intro	Within	Introspect
Intra	Within	Intracellular

Con/Com/Col/Cor/Co	With/Together	Consolidate
		Compound
Note: Com (before m, b, p)		Collision
Col (before l)		Correspond
Cor (before r)		Cooperate
Co (before a vowel)		
Over	Above	Overabundance
Ad	To/Toward	Adhesion
Epi	Upon	Epidermis
Mono	One	Monosaccharide
Of	Against	Offense

Some common roots which you should know are as follows:

Root	**Meaning**	**Example**
Capere	to take, seize	receptionist
(cep, cept, cap)		accept
Tenere	to hold	tenacious
(ten, tain)		maintain
		entertain
Mittere	to send	submit
(mit, mis)		missile
Ferre	to carry	fertile
(fer)		preferred stock
Stare	to stand	stability
(sis, sist, sta)		basis
		insist
Graph	to write	paragraph
(graphien)		geography
		biography
Legein	to study	biology
(logue, logy)		logic
Specere	to see	spectator
(spec)		speculate
Plicare	to fold	duplicate
(plic)		replicate
Tendere	to stretch	extend
(tend)		extensive
		pretend
Ducere	to lead	produce
(du, duce)		reduction
Popere	to place, put	deposit
(pos)		composition
Facere	to make, do	factory
(fic, fac)		facility
Scribere	to write	describe
(scrib, script)		inscription

3. In order to retain meanings for terms you've learned, you need to develop a systematic method of studying words found in your science textbooks:

a) Keep a vocabulary notebook or set of 3x5 index cards. In each entry (one entry per card), include the term, its pronunciation as found in a dictionary, and the word's common meaning. Also, write a sentence using the term in a science context.

```
                                                           FRONT

   dominant trait        (dŏm′ ə nənt trāt)

      Genetic trait which dominates or prevents the expression of the
      recessive trait.
```

```
                                                           BACK

      Although Mary's father had blue eyes, her eyes were brown
      because her mother's eyes carried the dominant trait brown.
```

b) Associate the term/word with a mental picture if possible.

c) Mark the term when you mark your textbook and include the term and its definition in your set of notes taken from the text. Reading and Marking Textbook Chapters and Taking Notes from Reading Assignments are explained as Skills 4 and 6 in this book. Explanations of these skills show you how to mark terms and definitions and how to include them in your notes.

Exercises

There are 9 exercises for Skill 3. To complete the first four exercises, you will need to use context clues and write the meaning of terms as well as use them in sentences. To complete the next four exercises, you will need to work with prefixes and roots using the lists on pages 96 and 97. For the last exercise, you will need to use a dictionary and create 3x5 vocabulary cards. You can check your work with the answer key at the back of the book. All words found in these exercises are drawn from the Sample Chapters in science or related science textbooks.

EXERCISE 1

Use context clues based on "definition" to determine the meaning of each italicized term below. Write the term and its meaning; then write a sentence of your own using the term in a science context.

1. *Digestion* is the process of breaking large molecules into smaller ones by chemical and physical means.

    a. Term:
    b. Meaning of the term:

    c. Complete sentence using the term:

2. *Pasteurization* is the process of heating and then rapidly cooling a substance to kill dangerous bacteria.

    a. Term:
    b. Meaning of the term:

    c. Complete sentence using the term:

3. *Evaporation* is the process whereby a liquid becomes a gas and likewise changes from the liquid to the gaseous state.

    a. Term:
    b. Meaning of the term:

    c. Complete sentence using the term:

4. *Oxidation*, the combination of a substance with oxygen, may occur at ordi-
    nary temperatures involving some substances such as yellow phosphorous or
    oily rags; these substances react at an ever-increasing rate as the temperature
    rises.

    a. Term:
    b. Meaning of the term:

    c. Complete sentence using the term:

5. *Kindling temperature* is the temperature at which a combustible substance
    bursts into flame.

    a. Term:
    b. Meaning of the term:

    c. Complete sentence using the term:

EXERCISE 2

Use context clues based on "examples" to determine the meaning of each italicized term below.
Write the term and the example(s) given. Then write a sentence of your own using the term in a
science context.

6. *Chain reaction* is a reaction that yields products that cause further reactions
    of the same kind and thus one which becomes self-sustaining. For example,
    the reaction between gaseous hydrogen and gaseous cholorine in sunlight is a
    reaction of this type.

    a. Term:
    b. Example(s):

    c. Complete sentence using the term:

7. A *solute* in a solution is the dissolved substance, the solid or gas which is dissolved in a liquid. For example, salt which is dissolved in a beaker of water is the solute.

    a. Term:
    b. Example(s):

    c. Complete sentence using the term:

8. In a solution, the medium in which the solute is dissolved is the *solvent*. For example, water is a most widely used solvent.

    a. Term:
    b. Example(s):

    c. Complete sentence using the term:

9. Each individual fluid has a characteristic amount of internal friction or resistance to flow that is known as *viscosity*. For example, viscosity of heavy lubricating oils is high, especially in comparison to liquids of low viscosity such as ether or benzene.

    a. Term:
    b. Example(s):

    c. Complete sentence using the term:

10. *Sublimation* occurs when substances under certain circumstances change directly from the solid to the vapor state and from the vapor state to the solid state. For example, "dry ice" (solid carbon dioxide) evaporates directly to gaseous carbon dioxide at temperatures above $-78.5°$ C, and does not pass through the liquid state.

   a. Term:

   b. Example(s):

   c. Complete sentence using the term:

EXERCISE 3

Use context clues based on "contrast" to determine the meaning of each italicized term below. Write the term and the term or phrase with opposite meaning. Then write a sentence of your own using the term in a science context.

11. He wanted to use *topsoil* for his houseplants; instead he used soil that contained little organic material and other substances and did not hold water and air well.

   a. Term:

   b. Term or phrase with opposite meaning:

   c. Complete sentence using the term:

12. The professor developed a *habit* of eating only healthy foods unlike his colleague who had to convince himself prior to each meal to eat only those foods high in nutritional value.

   a. Term:

   b. Term or phrase with opposite meaning:

   c. Complete sentence using the term:

13. Although thought to be *extinct*, the rare bird was found to be an existing form of life.

    a. Term:

    b. Term or phrase with opposite meaning:

    c. Complete sentence using the term:

14. Although said to be a *desert*, the arid region had rainfall of more than 25 cm per year and had vegetation that was continuous or narrowly spaced.

    a. Term:

    b. Term or phrase with opposite meaning:

    c. Complete sentence using the term:

15. Although he thought the substance was an *acid*, he learned that in solution it had a larger number of hydroxide ions than hydrogen ions.

    a. Term:

    b. Term or phrase with opposite meaning:

    c. Complete sentence using the term:

EXERCISE 4 ▬▬▬▬▬▬▬▬▬▬

Use context clues based on "additional information" to determine the meaning of each italicized term below. Write the term and the additional information provided. Then write a sentence of your own using the term in a science context.

16. Organisms as they are today are *adapted* to their environment enabling them to survive and reproduce. Three major types of adaptations are morphological, physiological and behavorial.

    a. Term:
    b. Additional Information:

    c. Complete sentence using the term:

17. As it is developing, a new organism is called an *embryo*. Embryos are live and have the same basic functions as any other living organism in that they secure food, obtain oxygen, rid themselves of wastes, and respond to their environment.

    a. Term:
    b. Additional Information:

    c. Complete sentence using the term:

18. In *external fertilization*, eggs are fertilized outside the female. External fertilization occurs in sponges, jellyfish, most worms, fish, and frogs.
    a. Term:
    b. Additional Information:

    c. Complete sentence using the term:

19. In *sexual reproduction*, there is a union (fusion) of two sets of DNA. One set of DNA comes from each parent.

    a. Term:

    b. Additional Information:

    c. Complete sentence using the term:

20. One form of *asexual reproduction* is vegetative propagation. In vegetative propagation, a new organism is produced from a nonsexual (vegetative) part of the parent organism as through fission in one-celled organisms including bacteria.

    a. Term:

    b. Additional Information:

    c. Complete sentence using the term:

EXERCISE 5 ▰▰▰▰▰▰▰▰▰▰▰▰▰▰▰▰▰▰▰▰▰▰▰▰▰▰▰▰▰▰▰▰▰▰

Using prefixes on pages 96-97, write a prefix in each blank below to correctly complete the word as defined. All words are found in science textbooks.

21. _____mune system: cells and tissues which identify and defend the body against foreign chemicals and organisms

22. _____vergence: evolutionary process in which distantly related species produce descendants which resemble each other

23. _____pulse: the product of the force and the time during which it acts

24. _____fraction of light: when light passes from one transparent material into another, part of the light is reflected at the surface and part of it passes into the new medium; the latter is bent and the light changes direction

25. _____clusion principle: no more than two electrons can be in any given quantum level; thus only two electrons can be in the lowest energy state

26. _____pression: type of stress that is caused when an object under tension is being stretched; however, when the two forces act toward each other, they act to shorten or compress the object

27. _____phase: period between mitoses during which chromosomes are replicated

28. _____composer: bacterium or other tiny living thing which breaks materials of once living things and their wastes into simpler forms so they can be used again

29. _____sociation: process in which the ions of ionic compounds separate in solution

30. _____saccharide: single sugar; basic building block for complex carbohydrates

EXERCISE 6 ▰▰▰▰▰▰▰▰▰▰▰▰▰▰▰▰▰▰▰

Using the root definitions on page 97, write the correct meaning of the terms below found in science texts.

	WORD	MEANING
31.	efficient	
32.	biology	
33.	replication	
34.	differentiation	
35.	reproduction	
36.	application	
37.	conductors	
38.	composition	
39.	spectrum	
40.	emission	

EXERCISE 7 ▰

Match each science term with its meaning. Write the letter for the correct meaning in the blank beside each term. Use the prefix and root lists on pages 96-97 to help you.

41. _____concussion

42. _____dehydration synthesis

43. _____refraction

44. _____resonance

45. _____coherence

46. _____Conservation
         (of energy)

47. _____inertia

48. _____impulse (of a force)

49. _____transistor

50. _____reaction

a. occurs when periodic impulses are given to an object at a frequency equal to one of its natural frequencies by oscillation

b. exists for two sources of waves if there is a fixed phase relationship between the waves they emit during the time the waves are being observed

c. brain bruise resulting from a severe fall or blow to the head

d. chemical reaction in which a large molecule is formed from smaller molecules by removing water

e. a homogeneous mixture or solid solution, usually of two or more metals; the atoms of one are replacing or occupying interstitial positions between the atoms of the other

f. the bending of a light beam when passing from one medium to another

g. apparent resistance which a body offers to changes in its state of motion

h. the total amount of energy in a system isolated from the rest of the universe always remains constant, although energy transformation from one form to another, including rest energy, may occur within the system

i. a response to a stimulus; a reverse or opposing action; a chemical change or transformation in which a substance decomposes, combines with other substances or interchanges constituents with other substances

j. transmits electricity in one direction and resists its flow in the opposite direction

k. the product of the force and the time during which it acts; a vector quantity having the direction of the force

EXERCISE 8 ▮

Some prefixes change their spelling for the sake of euphony. For example, "in" plus logical becomes illogical. Write the words below by joining the prefix to the main part of the word. Remember to spell the prefix correctly. Use the prefix list on pages 96-97 to help you.

51. con + lisions = _____
52. in + pedence = _____
53. con + nective = _____
54. in + migration = _____
55. con + ordination = _____
56. in + munological = _____
57. in + ternal = _____
58. con + parative = _____
59. con + roborate = _____
60. con + laborate = _____

EXERCISE 9 ▮

Write vocabulary card entries for each term listed below. On the front of each card, write the term, its pronunciation as found in the dictionary, and the word's common meaning. On the back of each card, write a sentence using the term in a science context.

61. Term: presuppose

front	
_____ _____	
term pronunciation	
_____	
_____	

Sentence	back
_____	
_____	
_____	
_____	
_____	

62. Term: evolution

front	
_____ _____	
term pronunciation	
_____	
_____	

Sentence	back
_____	
_____	
_____	
_____	
_____	

63. Term: ecology

front	
——————— ———————	
term pronunciation	
———————————————	
———————————————	

Sentence	back
———————————————	
———————————————	
———————————————	
———————————————	
———————————————	

64. Term: subspecies

front	
——————— ———————	
term pronunciation	
———————————————	
———————————————	

Sentence	back
———————————————	
———————————————	
———————————————	
———————————————	
———————————————	

65. Term: extrapolate

front	
——————— ———————	
term pronunciation	
———————————————	
———————————————	

Sentence	back
———————————————	
———————————————	
———————————————	
———————————————	
———————————————	

66. Term: erode

front	
——————— ———————	
term pronunciation	
———————————————	
———————————————	

Sentence	back
———————————————	
———————————————	
———————————————	
———————————————	
———————————————	

67. Term: mutate

	front
_____ _____	
term pronunciation	
_____	
_____	

Sentence	back
_____	
_____	
_____	
_____	
_____	

68. Term: transpose

	front
_____ _____	
term pronunciation	
_____	
_____	

Sentence	back
_____	
_____	
_____	
_____	
_____	

69. Term: monogamy

	front
_____ _____	
term pronunciation	
_____	
_____	

Sentence	back
_____	
_____	
_____	
_____	
_____	

70. Term: dispersion

	front
_____ _____	
term pronunciation	
_____	
_____	

Sentence	back
_____	
_____	
_____	
_____	
_____	

Summary

1. In addition to reading a wide range of materials in the field, there are direct steps you may take in order to increase your vocabulary.

2. One important step to take when attempting to determine the meaning of new terms is to use context clues. Four common types of context clues are definition, examples, contrast, and additional information. Definitions are statements directly explaining the meaning of specific terms. Examples are real life or commonly known instances of the term provided to clarify a term's meaning. Contrast refers to an author's use of a more clearly understood term or phrase opposite in meaning to the term he/she is explaining. Additional information is provided to help clarify the meaning of a term after the more basic definition or some explanation has been provided.

3. Another step to take to unlock the meaning of terms found in science material is to apply the meanings of commonly used prefixes and roots and to apply that knowledge when determining the meaning of words.

4. In order to remember terms and their meanings and to be able to use them effectively, you should develop a systematic method of studying words. An effective system can be summarized as follows:
    a. Keep a vocabulary notebook or set of 3x5 index cards to include the terms to be remembered, their dictionary pronunciations and meanings, and sentences of your own using the terms.
    b. Associate the terms with mental pictures if possible.
    c. Mark the terms as you mark your textbook and include them in your notes as explained in Skills 4 and 6 in this book.

Additional Information

As demonstrated in the sample chapters in this book, science textbooks are replete with terms typically explained through definition, examples, contrast and additional information. It is important to remember that numerous terms found in science material may be found in other materials too. In a science context, however, the words have very specific meanings which you need to remember. It is important to learn definitely the meanings of the terms as used in the areas of science.

Science students may need to know also metric unit prefixes and their definitions. For reference, see below:

giga	1,000,000,000 =	10^9
mega	1,000,000 =	10^6
kilo	1,000 =	10^3
hecto	100 =	10^2
deka	10 =	10^1
deci	0.1 =	10^{-1}
centi	.01 =	10^{-2}
milli	.001 =	10^{-3}
micro	.000001 =	10^{-6}
nano	.000000001 =	10^{-9}

4

Reading and Marking Chapters

When you read a chapter in a science textbook, the quickest way to focus on the information that you need to learn is to mark the chapter as you read. Marking each section of the chapter as you read it can help you understand and remember what you read.

APPLYING THIS SKILL

To practice applying this skill, you need:

1. This book
2. A <u>colored</u> pen or pencil

Take these items to a work area. Read the review section to make sure you understand the information you need in order to apply this skill. Then complete the exercises for Skill 4.

PERFORMANCE OBJECTIVE

When you have completed Skill 4, you will be able to identify and mark important information when you read chapters in science textbooks, by using four signals: headings, key terms, lists, and illustrations.

Reading & Marking Textbook Chapters

When you read a chapter in a textbook, the quickest way to focus on the information that you need to learn is to mark the information. Don't mark too much, however, or you may confuse important information and how it's organized with less important information.

At the end of this review is a sample section from a chapter in a biology textbook which has been marked for you. You may look at it in relation to the points listed below regarding how to mark textbook material. A colored pen was not used here, but it is a good idea to use one or more when you mark the information you need to remember.

Let's review what you need to know in order to apply this skill.

1. When you read and mark a chapter, four signals of important information are:
   a. headings
   b. key terms
   c. lists
   d. illustrations

2. To mark important information signaled by headings:
   a. Underline all the headings
   b. If a heading has one or more paragraphs right under it, turn the heading into a question.
   c. Write the question word (who, how, or what) next to the heading.
   d. Read the section to find the answer to your question.
   e. Underline the answer and put ANS in the margin next to it.

3. To mark key terms and definitions:
   a. Underline a key term and its definition.
   b. Put DEF in the margin next to the definition.
   c. If the DEF is also an answer (ANS) to the question word next to the heading, write ANS-DEF in the margin.

   In science textbooks, you will find many key terms defined in each chapter. The terms may be printed in italics, but their definitions which follow usually appear in regular print.

4. To mark important information signaled by lists:
   a. Circle the words that tell you what a list is about.
   b. Number the items in the list if they do not already have numbers and letters.
   c. If the list is also an answer (ANS) to the question word next to the heading, write ANS in the margin next to the words which introduce the list. In this case, you would not have to underline the answer in the text.

Since science textbooks provide you with many facts, formulas, and procedures, their authors frequently present information in the form of lists. The items in a list are sometimes already numbered. If they are, you will not need to number them. Remember, however, to circle the words which tell you what the list is about. These words are usually in the sentence which comes right before the items listed.

5. To mark illustrations:
   a. Underline the title of the illustration, if it has one. Most illustrations in science textbooks do have titles. Furthermore, two or three additional sentences often follow the title, more clearly explaining the illustration.
   b. If an illustration has no title, use the explanation of the illustration to give it a title. Explanations are usually given on the same page as the illustration.
   c. Write your title near the illustration.

In science textbooks you will find many illustrations. Be prepared to underline numerous titles of illustrations as you mark your textbook.

6. Review each textbook section after you read and mark it. Go over information that you have marked.

7. Finally, review the whole chapter when you have finished reading and marking it. Go over everything you have marked. Answer the review questions and complete any exercises/problems listed in the chapter closing. If there is anything in the chapter that is still not clear to you, write down your questions and discuss them with your instructor.

When reviewing a chapter in a science textbook, you should also complete exercises and problems in the related section of the study guide if one accompanies your textbook. In study guides there are usually true-false, multiple-choice, short answer, and essay questions about the material in the chapter. Problems requiring mathematical computation and application of formulas are included in some study guides too. Answering these questions and solving the problems, in addition to those included in the chapter closing, will further assist you in learning and remembering important points and procedures covered in the chapter.

Look at the sample section below which has been marked for you.

Biology — Classification System

What?   *Taxonomy*. Biologists have long been interested in the problem of classifying organisms. Many different kinds of living things are on earth today. There are nearly two million species of organisms. The study of the classification of plants
ANS- DEF   and animals is called *taxonomy*. Taxonomy, the oldest branch of biology, has required much scientiic effort.

What?   *Earlier classification systems*. There were two classification systems used many
ANS   years ago. One system, developed by Aristotle, separated living things into two major groups — plants and animals. Animals were classified according to where they lived. Plants were classified according to structure. [2] A second classification system, developed by Linnaeus in the 1700's, separated living organisms according to their structural features. Each type of organism was considered a distinct species. Organisms with the same set of features were classified as members of the same species.

In Linnaeus' system, binomial nomenclature was used for naming organisms.
DEF   *Binomial nomenclature* is a two-term naming system. Two words are used to name an organism. [1] The first word identifies the genus to which an organism belongs.
DEF   *Genus* is a category consisting of closely related species of organisms. [2] The second
DEF   word represents the particular species. *Species* is a group of related organisms which can interbreed and which reproduce in a different way than organisms of other groups. Genus-species names are still used by scientists all over the world.

What?   *Today's classification system*. Homologous structures serve as the basis for
ANS   classifying organisms today. Organs or parts in different types of organisms which seem to be related are said to be homologous. *Homology* is correspondence in
DEF   structure and function indicating common ancestry among organisms. When homologous structures serve as the basis for classification, seven groups are used leading from general to specific: kingdom, phylum, class, order, family, genus, and
[7] species.
See Figure 1-1.

Figure 1-1. **The Classification of a Swallowtail Butterfly**
From Kingdom Down to Species

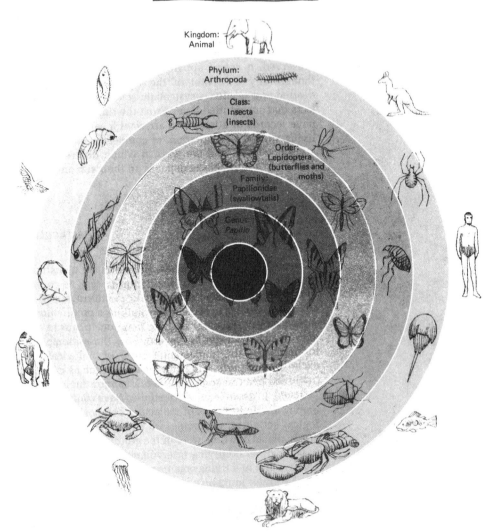

From: Thomas C. Emmel, **Worlds Within Worlds: An Introduction
to Biology** (Orlando: Harcourt Brace Jovanovich, 1977), p. 264.
Reprinted by permission of the publisher.

Here you see the classification of a swallowtail butterfly from kingdom to species. Classification can be more complex than this. Each of these major groups may be subdivided further.

EXERCISES ■■■■■■■■■■■■■

There are 9 exercises for Skill 4. For each exercise, you will read and mark a section from one of the sample chapters you surveyed in Skill 2. The textbook sections have been reprinted on the following pages, so you can mark them. Each section contains one or more of the four signals of important information. Read each section, looking for the signals, and mark the important information.

Exercises 1-3 are sections from biology textbooks. Exercises 4-6 deal with physics. Exercises 7-9 are about chemistry. You may complete as many exercises as you need, and you may work on the exercises in any order. You can check your answers with the answer key at the back of this book.

EXERCISE 1 ■■■■■■■■■■■■■

From the chapter "Algae."

Blue-Green Algae

Basis for color. Blue-green algae contain blue pigment. In combination with chlorophyll, which is green in color, the blue pigment gives algae their blue-green color. Some kinds of blue-green algae also contain a red pigment. Red pigment, in combination with other pigments, creates a near black color in some "blue-green" algae.

Cell structure. Blue-green algae are organized in a simple manner. They are the simplest in organization of all algae in several ways. First, there isn't a definite nucleus in a single cell. As a result, DNA is scattered throughout the cell. *DNA* is the material that carries the genetic information for reproduction. Second, chlorophyll in blue-green algae is usually attached to membranes. There are no chloroplasts to contain the chlorophyll. Third, single cells are often arranged in chains or filaments as shown in Figure 12-1. Fourth, there seems to be little division of labor among the cells. Each cell is like every other cell of the group. Fifth, groups of cells are often enclosed within a protective jellylike layer.

Figure 12-1. **Blue-Green Algae**

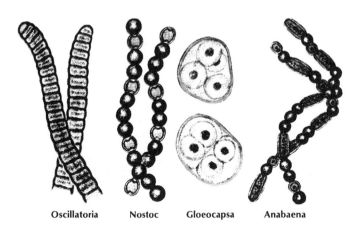

Oscillatoria   Nostoc   Gloeocapsa   Anabaena

From: Raymond F. Oram, Paul J. Hummer, Jr., and Robert C. Smoot, **Biology: Living Systems** (Columbus, Ohio: Charles Merrill Pub. Co., 1976), p. 314. Reprinted by permission of the publisher.

How blue-green algae reproduce. Blue-green algae reproduce asexually (without mating) through division of a single cell. New cells may be formed along a chain or filament through simple fission. *Fission* is reproduction in which a one-celled organism divides into two one-celled organisms. Occasionally, filaments are broken apart. Each section produces a new series of cells by fission. One kind of blue-green algae called Anabaena produces specialized cells that act as spores. These spores can develop into new filaments.

Habitat. Blue-green algae are commonly found in ponds, streams, and moist places on land. During hot summer months they multiply rapidly. As a result, public water supplies must be treated regularly during this season. Treatment is necessary to prevent rapid reproduction of blue-green algae and other microorganisms.

EXERCISE 2

From the chapter "Protozoa."

Amoebae

Forming pseudopodia. One of the simplest protozoans is the amoeba. Amoebae are unique in how they move. To move, they develop pseudopodia. *Pseudopodia* are false feet formed by flexible plasma membranes and cytoplasm. The formation of pseudopodia involves certain stages. Initially, the cytoplasm is in a fixed, gelatinous state. Then, chemical changes cause part of the cytoplasm to become a fluid. This fluid, which is like a true solution, is known as a sol. Cytoplasm in a sol condition can flow freely. Next, cytoplasm flows into the cell membrane, extending it to form a pseudopodium. Then this cytoplasm changes back to a gel. Finally, other areas of the cytoplasm undergo the same sol-gel transformation forming new pseudopodia. New pseudopodia are formed and others disappear. In this way, the shape of amoebae changes constantly. See Figure 13-1.

Figure 13-1. **Movement of the Amoeba**

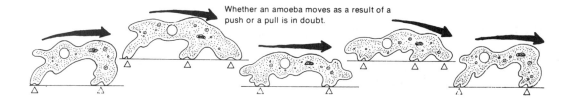

Whether an amoeba moves as a result of a push or a pull is in doubt.

From: Raymond F. Oram, Paul J. Hummer, Jr., and Robert C. Smoot, **Biology: Living Systems** (Columbus, Ohio: Charles Merrill Pub. Co., 1976), p. 327. Reprinted by permission of the publisher.

Food-getting structures. Pseudopodia are used in getting food as well as in moving about. Pseudopodia engulf particles of food. The food particles enter the cytoplasm where they are digested. You can see the food vacuoles in which digestion takes place. See Figure 13-2.

Figure 13-2. **Food-Getting Structures of the Amoeba**

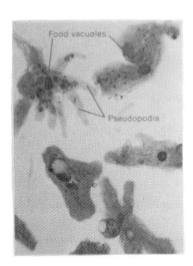

Habitat. The kind of amoeba studied in the laboratory is commonly found in ponds and other bodies of water. This type of amoeba is harmless. Some amoebae, however, are found in contaminated water. They can cause disease and are quite dangerous to humans.

EXERCISE 3

From the chapter "Origin of Genetics."

Mendel's Initial Experiment

Purpose. Mendel experimented with pea plants to discover how traits are transmitted from parent to offspring. He wanted to arrive at a set of rules for the transmission of traits. He had already made certain observations about pea plants. First, he had seen that traits or visible characteristics are hereditary. *Hereditary* means that traits are passed on from generation to generation. Second, he had seen that many traits exist in one of two possible forms. For example, pea seeds are either round or wrinkled. The stems of pea plants are either tall or short.

Plants used in the experiment. For his experiment, Mendel used pure pea plants. *Pure plants* are those which after many generations of offspring have the same features as the parents. Pea plants reproduce sexually. There are both male and female sex organs in the same flower. Normally, male gametes (sex cells or pollen) fertilize eggs of the same flower. Pollen from the anther fertilizes an egg in an ovule. See Figure 11-1. The ovule becomes the seed. The ovary develops into a pod which protects the seeds. After many generations this process results in offspring with the same features as the parents or pure pea plants.

Figure 11-1. **Sexual Reproduction of Pea Plants**

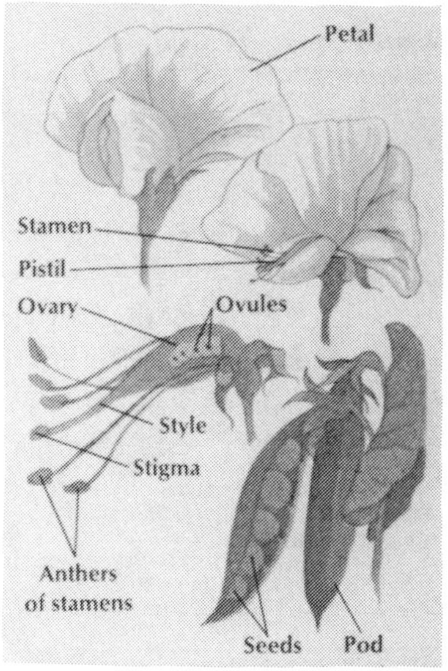

From: Raymond F. Oram, Paul J. Hummer, Jr., and Robert C. Smoot, **Biology: Living Systems** (Columbus, Ohio: Charles Merrill Pub. Co., 1976), p. 146. Reprinted by permission of the publisher.

Mendel's accomplishment. Mendel crossed pure plants having a particular trait with pure plants having the opposite trait. More specifically, he crossed pea plants which produced round seeds with pea plants which produced wrinkled seeds. To do this, he transferred the pollen of a plant with one trait to plants with the opposite trait, or opposite form of the trait. Pollen was obtained, therefore, from round-seeded plants and transferred to wrinkle-seeded plants. The opposite procedure was also performed.

Results. Mendel found that in every case the parental cross (P) yielded offspring with round seeds only. The offspring of a parental cross are called the *first filial*, or F_1, *genera-tion.* There were no wrinkle-seeded plants in the F_1 generation.

Law of dominance. Mendel concluded that for each trait there is one form which "domi-nates" the other. A *dominant trait* is the trait which appears exclusively in the F_1 genera-tion. For example, the trait for round seeds is the dominant trait. It "dominates" the trait for wrinkled seeds. The trait which disappears in the F_1 generation is called a *recessive* trait. In this case, the trait for wrinkled seeds is the recessive trait. Mendel generalized this result and formulated the law of dominance. The Law of Dominance states that one trait, the dominant trait, dominates or prevents the expression of the recessive trait.

EXERCISE 4 ▮▮▮▮▮▮▮▮▮▮▮▮▮▮▮▮▮▮▮▮▮▮▮

From the chapter "Speed, Velocity, and Acceleration."

Acceleration

Acceleration response. Every driver knows that he must press down the accelerator pedal to increase the speed of his car. All cars do not, however, respond in the same way to a depression of an accelerator. Some will change speed much faster than others. The *acceleration response* refers to the time it takes to achieve a certain speed starting from rest. If you compare the acceleration of two cars, the one reaching a specified speed in the least time has the greater acceleration. Acceleration can be expressed by the formula:

$$\text{acceleration} = \frac{\text{change in speed}}{\text{time required for change to occur}}$$

Direction and acceleration. Direction can be part of acceleration. Acceleration then refers to a change in velocity rather than speed. You can say then that an object can be accelerated by changing either (or both) the speed and direction.

Consider an example. A car travels at a constant speed of 50 miles/hour. It is traveling on a circular track. The car is accelerating because its direction is constantly changing. See Figure 2-1.

Figure 2-1. **Direction and Acceleration**

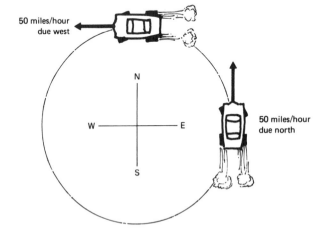

From: Joseph Priest, **Energy for a Technical Society,** ©1975, Addison-Wesley Pub. Co., Inc., Reading, Massachusettes, p. 50. Reprinted with permission.

EXERCISE 5 ▮▮▮▮▮▮▮▮▮▮▮▮▮▮▮▮▮▮▮▮▮▮▮▮

From the chapter "Newton's Three Laws of Motion."

Newton's Second Law of Motion

What causes resistance? Suppose that you kicked different kinds of objects on the floor with the same force. Would all of the objects achieve the same acceleration? Surely they would not. Every object offers some resistance to being put into motion. Resistance is caused by the property of mass known as inertia. *Inertia* is the property or tendency of a body to resist any change in its state of motion, be it starting, stopping, or changing it from a straight-line path. The greater the weight, the greater the mass, the greater the inertia, the greater is the resistance to motion.

Assign a number to measure mass. A simple balance can be used to assign a number to measure mass. It compares the unknown mass with the known mass. See Figure 4-1. The units used are kilograms. The symbol for kilogram is kg.

Figure 4-1. **Measuring Mass**

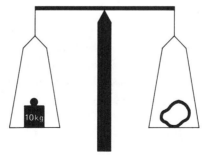

From: Joseph Priest, **Energy for a Technical Society**, ©1975, Addison-Wesley Pub. Co., Inc., Reading, Massachusettes, p. 31. Reprinted with permission.

Second law, force, and equations. Newton's second law quantifies these ideas discussed above. The second law states: "Acceleration experienced by an object is proportional to the net force acting on it and is inversely proportional to its mass." It then defines *force* as the amount of acceleration given a mass. In equation form you can say:

$$\text{Acceleration} = \frac{\text{net force}}{\text{mass}}$$

$$\mathbf{a = F/m} \ \text{ or } \ \mathbf{F = ma}$$

The smaller the mass of an object acted upon by a given force, the greater its acceleration and hence, final velocity. See Figure 4-2.

Figure 4-2. **Mass and Acceleration**

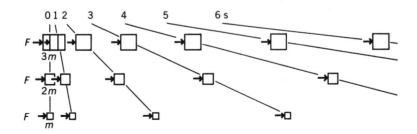

When the same force is applied to bodies of different masses, the resulting accelerations are inversely proportional to the masses.
Succesive positions of blocks of mass, **m, 2m,** and **3m,** are shown at 1-s intervals while identical forces of **F** are applied.

From: Arthur Beiser, **Physics,** 2nd Edition (Menlo Park: CA: The Benjamin/Cummings Pub. Co., 1978), p. 67. Reprinted by permission of the publisher.

The Newton. It is convenient to use a special unit for force. In the SI system of units (Systeme International Version of the Metric System), the unit used for force is the newton. Newton is abbreviated: *N.* A *newton (N)* is that force which, when applied to a 1 kg mass, gives it an acceleration of 1 m/s². Here, as in Figure 4-2, "s" refers to seconds and "m" refers to mass.

EXERCISE 6

From the chapter "Work."

Work from the Physics Standpoint

Work and its equation. From the physics standpoint, physical effort directed toward the production of something is supplied by a force. Work is said to be done when a force acts on a mass while it moves through some distance. A physical quantity called *work* can be defined as follows:

The work done by a force acting on an object which moves in the same direction as the force is equal to the magnitude of the force multipled by the distance through which the force acts.

The definition of work in equation form reads:

$$Work = force \times distance$$

$$W = Fs$$

Measuring work. Work can be measured in units of foot pounds and joules. The foot pound (ft. lb.) is the unit of work in the British system of units. One *foot pound* is equal to the work done by a force of one pound acting through a distance of one foot.

For example, a man may push a refrigerator. The amount of work he does is equal to the magnitude or amount of force he applies times the distance he moves the refrigerator. He may push the refrigerator 10 feet with a force of 20 pounds. If he does this, he does 20 lb. x 10 ft. = 200 foot pounds of work.

The joule (J) is the unit of work in a more recent system of units, the Systeme International (SI), the current version of the metric system. One *joule* is equal to the work done by a force of 1N (newton) acting through a distance of 1m(meter), i.e. 1J = 1N x m. A *newton* is a metric unit of force that is equivalent to a force of about 1/4 pound.

It's important to realize, however, that a force may be exerted and yet no work may be done. This happens if the force does not act through a distance. For example, a man pushing against a brick wall does no work on the wall if the wall does not move. The product of force and distance is zero because the distance is zero. Only a force that gives rise to motion is doing work.

Positive work and negative work. It is important to recognize that a force can act in one of two ways. First, it can act in the same direction as the mass is moving (positive work). Second, force can act opposite to the direction in which the mass is moving (negative work). In both cases work is being done. (See Figure 7-1, parts a and b)

Figure 7-1 **Positive Work and Negative Work**

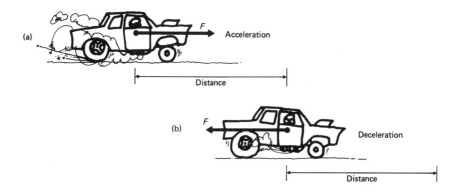

(a)   Force acts in the same direction that the car
       moves. This tends to increase the car's speed.
       This is a case of positive work.

(b)   Force acts in a direction opposite to that in
       which the car moves. This tends to decrease
       its speed. This is a case of negative work.

From: Joseph Priest, **Energy for a Technical Society**, ©1975,
Addison-Wesley Pub. Co., Inc., Reading, Massachusettes, p. 34.
Reprinted with permission.

Net work. The *net work* on an object is the algebraic sum of the positive and negative works
done by each force acting on the object. Suppose that two forces of +100 N and -50N
caused an object to be moved a distance of 2 meters. Then the work done to the two forces
would be:

$$(+100) \ (+2) \ = \ +200J$$
$$\text{and}$$
$$(-50) \ (+2) \ = \ -100J$$

The net work would be +200 - 100 = +100J

EXERCISE 7 ███████████████████████████████████████

From the chapter "Atoms and Molecules."

Dalton's Assumptions

Assumptions. Dalton made two assumptions about atoms and molecules. First, Dalton hypothesized that all the atoms of a particular element are alike and have the same mass. An *atom* is the smallest particle of an element. An *element* is a substance composed of only one type of element; it cannot be broken into other substances by chemical means. Second, Dalton assumed that chemical combination involves chemical bonding. This bonding of atoms of each of the combining elements forms molecules of the resulting compound. Bonding occurs in a predictable manner.

Law of Conservation of Mass. Dalton's atomic theory helped explain the Law of Conservation of Mass. The Law of Conservation of Mass tells us that there is no detectable gain or loss of mass during any physical or chemical change. Note that the Law of Conservation of Mass is applied to chemical reactions as well as to physical transformations. Atoms involved in both kinds of processes keep their identities. In a chemical reaction, the molecules of the products contain the same atoms that were initially present in the molecules of the reactants. When charcoal burns, for example, there are two reactants. One is an atom of carbon. The second is a molecule of oxygen. One atom of carbon combines with one molecule of oxygen. The product or result is a carbon dioxide molecule. Each individual atom retains its original mass. The sum of masses of the molecules of the product is the sum of the masses of the molecules of the reactants.

EXERCISE 8 ████████████████████████████████

From the chapter "Periodic Table of Elements."

Groups

Vertical columns. The vertical columns in the table are chemical groups. The elements in any one column therefore make up a group.

Elements in a group or column. Elements in a group are a family of elements having similar properties. They're more like each other than they're like elements of other groups.

Additional points can be made about elements in the groups which are designated by Roman numerals in two series, A and B. First, the physical and chemical properties of elements in a group quite regularly change in qualities with increasing atomic number. For example, the melting points (Celsius) of the alkali metals are $_3$Li, 180°; $_{11}$Na, 98°; $_{19}$K, 64 °; $_{37}$Rb, 39°, $_{55}$Cs, 29°. See Figure 5-2.

Figure 5-2. **Melting Points of the Alkali Metals**

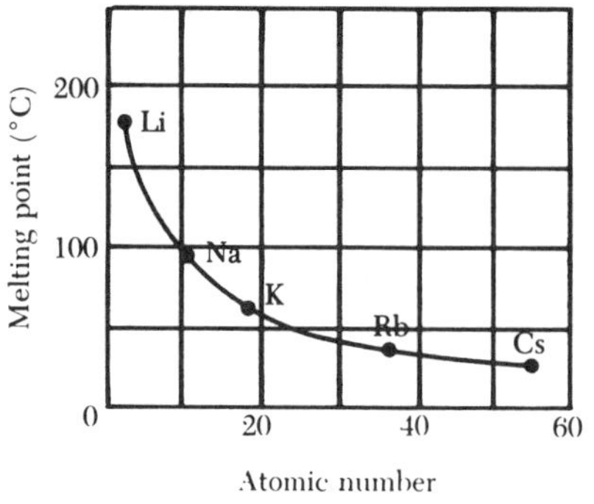

From: Lawrence P. Eblin, **The Elements of Chemistry,** 2nd Edition (Orlando: Harcourt Brace Jovanovich, 1970), p. 67. Reprinted by permission of the publisher.

Second, elements in the same group form compounds with similar formulas. For example, formulas for oxides of Group VB elements are: V_2O_5, Nb_2O_5, Ta_2O_5. Third, in the A groups, metallic properties are accentuated with an increase in the atomic number. Fourth, in the A group, nonmetallic properties diminish as the atomic number increases. Fifth, the first member of a group is often more different from the other members of the group than the other members are different from each other. For example, fluorine does not resemble chlorine as much as chlorine resembles bromine. Also boron is quite different from the other elements of Group IIIA. It is the only nonmetal in the group. Sixth, the members of an A group somewhat resemble members of the corresponding B group. For example, consider manganese in Group VIIB and chlorine in Group VIIA. They both form acids with similar formulas and properties.

Group names. Certain groups have been given names. Group IA metals (excluding hydrogen) are called alkali metals. Group IIA metals are the alkaline earth metals. On the other side of the table, Group VIIA are the halogens. Group VIIIA are the noble gases.

What location tells us. Where an element is located in the table tells us its physical and chemical properties. The elements on the left side, for example, are strongly metallic. The elements on the right side are nonmetallic. Then, some elements are between the metals and nonmetals. They are sometimes called metalloids. For example, consider arsenic and selenium. They're sold under the labels, "arsenic metal" and "selenium metal." Chemists would agree, however, that these elements should be classified as nonmetals. Remember also that what is true about one member of a group is usually true about the others in the same group.

EXERCISE 9

From the chapter "Relationship of Volume and Pressure of Gases."

Early Experimentation

Experiment. Sir Robert Boyle tried to discover if the volume and pressure of a gas were related. He used an apparatus with gas and mercury to measure the relationship. See Figure 7-1.

Figure 7-1. **Apparatus for Determining Relationship of Volume and Pressure of Gases**

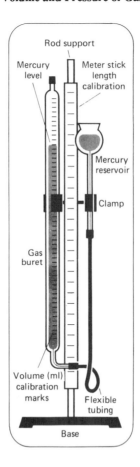

From: Richard M. McGurdy, **Qualities and Quantities: Preparation for College Chemistry** (Orlando: Harcourt Brace Jovanovich, 1975), p. 65. Reprinted by permission of the publisher.

A sample of gas (e.g., air, nitrogen, or carbon dioxide) was trapped in the apparatus. It was trapped over mercury in a closed device used to measure volume. The mercury acted as a piston to change the volume of the gas when the mercury reservoir was raised or lowered. The surface level of the mercury was related to the amount of pressure exerted.

Result. Boyle found that for gases, as the pressure increased the volume was decreased. Volume was decreased in proportion to the amount of pressure increase. Using a constant amount of gas at a constant temperature, he was able to show this relationship. Look at the figures in Table 7-1. Notice the definition of torr in Table 7-1 also.

Table 7-1. **Pressure versus Volume (at constant temperature)**

Pressure (torr)	Volume (ml)	PV Product
950	30.3	2.88×10^4
750	38.4	2.88×10^4
500	57.6	2.88×10^4
350	82.3	2.88×10^4
300	96.0	2.88×10^4

A *torr* is the pressure exerted by a column of mercury one millimeter high.

From: Richard M. McGurdy, **Qualities and Quantities: Preparation for College Chemistry** (Orlando: Harcourt Brace Jovanovich, 1975), p. 65. Reprinted by permission of the publisher.

These figures can be placed on a graph. The curve in the graph shows an inverse relationship. As pressure increases, volume decreases. Look at the solid line in Figure 7-2 on the following page.

Figure 7-2. **Pressure and Volume of Gases**

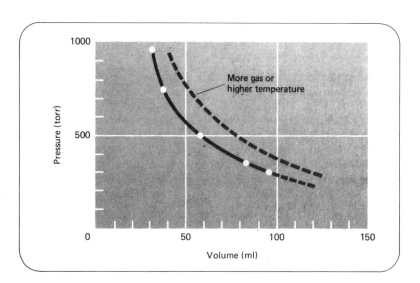

From: Richard M. McGurdy, **Qualities and Quantities: Preparation for College Chemistry** (Orlando: Harcourt Brace Jovanovich, 1975), p. 66. Reprinted by permission of the publisher.

Equation for relationship. The equation for the relationship between volume and pressure is $V = k/P$. V is volume. P is pressure. k is a proportionality constant. This constant, k, depends on the temperature and the amount of gas. As the pressure increases, the volume decreases. The fraction k/P gets smaller. It never becomes zero or negative, however. As the pressure decreases the volume increases. Volume becomes very great as the pressure nears zero.

A sample with more gas or higher temperature could be used too. It would result in a similar curve (dashed line) further from the axis. See Figure 7-2.

SUMMARY

1. A good way to help you understand and remember what you read when you study a textbook chapter is to mark the important information as you read, preferably with a colored pen/pencil.

2. You can use four signals to help identify and mark important information:
    a. headings
    b. key terms
    c. lists
    d. illustrations

3. A section heading usually refers to the most important information covered in the section. To mark a heading and the section that follows it:
    a. Underline the heading.
    b. Ask yourself a question about the heading, using one of the question words: Who?, What?, or How? Write the question word near the heading.
    c. Ask yourself, "What is important about this subject?"
    d. Read the section, looking for answers to your question. You may need to read through the whole section before you mark anything.
    e. Underline the answers to your question and write ANS in the margin where the answers appear.

4. Key terms are a special vocabulary of important words and terms in a subject. You should learn key terms and understand their definitions as you read. To mark key terms:
    a. Watch for key terms in boldface or italic type as you read.
    b. Underline the key terms and their definitions.
    c. Write DEF in the margin next to the definitions.

5. Lists present related information in a concise way. To mark a list:
    a. Circle the words that introduce and explain the list.
    b. If the items in the list do not already have numbers or letters, number them.

6. Illustrations often present important information in picture form. To mark an illustration:
    a. Underline the title, if there is one.
    b. If an illustration has no title, read the explanation and make up your own title. Write the title near the illustration.

7. Remember that you will often need to read through each short section of text first, and then decide how to mark it.

5

Using Maps, Diagrams Graphs, and Tables

In science textbooks, authors frequently use figures and tables to help you picture the information. Diagrams and graphs are the kinds of figures most often used. Tables, of course, are used too. Studying the figures and tables as you read the text of a chapter can help you understand the information more clearly and remember it more easily.

APPLYING THIS SKILL

To practice applying this skill, you need:

1. This book
2. A pencil

Take these items to a work area. Read the review section to make sure you understand the information you need in order to apply this skill. Then complete the exercises for Skill 5.

PERFORMANCE OBJECTIVE

When you have completed Skill 5, you will be able to interpret information represented in diagrams, graphs, and tables in science textbooks.

Using Maps, Diagrams, Graphs, & Tables

In science textbooks authors use many illustrations to help you picture the information. Often diagrams, graphs, and tables are the kinds of figures/illustrations used. They make the written ideas or facts clearer. They are sometimes used to present information which is referred to, but not thoroughly discussed in a chapter. You need to study these illustrations to get information from them. Understanding the information they contain is often required for clear comprehension of the chapter. Usually, the diagrams, graphs, and tables are numbered and titled. Titles which often appear in large print tell you what figures and tables are about. Always read the titles and any additional explanations for this information. They may appear at the top, bottom, or side of the figures and tables. The source of information in a figure or table may be given too. The source tells you where the information came from and when it was gathered. It may tell you who gathered the information, as well.

Authors of science textbooks almost always refer you to the illustrations by stating "See Table ___ " for tables, or "See Figure___" for graphs and diagrams within the appropriate sections of the chapter. As you read a section, study an illustration as soon as the author refers you to it. Then the meaning of the part you've just read and the part which follows will become clearer to you.

During lectures, instructors of science courses may use illustrations in transparencies, slides or large charts similar to those found in science textbooks. If you are quite familiar with interpreting information found in your textbook illustrations, understanding those used during lectures will be much easier.

Let's review now what you need to know in order to apply this skill to specific figures and tables.

1. A diagram is a drawing or sketch. Diagrams are used to help you picture an object, a series of events, a process, or the organization and classification of information. When you study a diagram, look for a title or explanation and any labels that help make the diagram clearer. Examine the parts of the diagram as you read the explanation provided.

2. An author uses a graph to help you compare sizes or amounts. When you study a graph, look for a title or explanation. There are different kinds of graphs. A circle graph is often called a "pie graph." When you study a circle graph, look at the entire circle, how a quantity is divided into portions, and any labels that identify those portions. The entire circle stands for the whole amount being examined. A circle graph shows how the whole amount of something is divided into portions. A bar graph contains a series of bars to help you compare several quantities or amounts. The bars may be either vertical or horizontal. When you study a bar graph, look at the vertical and horizontal scales to see what the bars represent. One scale will identify the bars; the other will act like a ruler used to measure the length of the bars.

Use the scales to see what quantity each bar represents. Then compare the bars to one another. A line graph is most often used to show how some quantity increases or decreases over time. There can be one line on a graph representing one quantity, or there can be several lines representing different quantities. When you study a line graph, look at the vertical scale and the horizontal scale to see what the line(s) represents. Use the scales to see what different points on the line stand for. If the graph has more than one line, compare the lines to one another. Also, look at the direction of a line(s). An upward sloping line usually represents an increase. A downward sloping line often shows a decrease. After observing the slope of the line(s), look for a pattern of increases or decreases.

3. Tables are used to present a large amount of detailed information in a short, easy-to-read form. They usually contain numerical or statistical information. Columns in a table run from top to bottom. The column headings tell you what information is found below them. Rows in tables run from left to right. To read a table, you look at the column headings first. Then you read across a row, looking at the information in each column. When you read across a row, you should compare the information in one column to the information in the other columns. Or you can read down a column and compare the different entries to each other. Most tables are organized so that you can compare the information in various ways. Sometimes notes are provided below a table. If notes are apparent, study them to get a better understanding of the entries in the table.

EXERCISES

There are 9 exercises for Skill 5. In each exercise, you will answer questions about a diagram, graph, or table in one of the sample chapters.

The directions at the start of each exercise tell you which diagram, graph, or table to use. Be sure to find the correct figure or table before you begin answering the questions. Write your answers in the spaces provided in this book. You can check your answers with the answer key at the back of this book.

EXERCISE 1 ▉▉▉▉▉▉▉▉▉▉▉▉▉▉▉▉▉▉▉▉▉▉▉▉▉

Use Figure 11-1 on page 224 in this book to answer Questions 1-4.

1.   What is the title of this diagram?

2.   According to the diagram, what outer part of the plant surrounds the reproductive organs?

3.   What is the name of the part that contains the ovules?

4.   What is the name of the part that appears at the outer end of the style?

EXERCISE 2 ▉▉▉▉▉▉▉▉▉▉▉▉▉▉▉▉▉▉▉▉▉▉▉▉▉

Use Figure 4-3 on page 234 in this book to answer Questions 5-7.

5.   What is the title of this diagram?

6.   What happens when the moving car exerts a force on the auto that is initially at rest?

7.   What do you think would happen if there were another stationary car in front of the one that was hit?

EXERCISE 3 ▉▉▉▉▉▉▉▉▉▉▉▉▉▉▉▉▉▉▉▉▉▉▉▉▉

Use Figure 7-2 on page 240 in this book to answer Questions 8-12.

8.   What is the title of this diagram?

9.   What letter is used to stand for "work"?

10.   What letter is used to stand for "mass"?

11.   What force does the work when an object falls to the ground?

12.   What can be said about the amount of work done by gravity when an object is moved parallel to the earth's surface?

EXERCISE 4 ▮▮▮▮▮▮▮▮▮▮▮▮▮▮▮▮▮▮▮▮▮▮▮

Use Figure 3-1 on page 244 in this book to answer Questions 13-16.

13. What is the title of this diagram?

14. Based on this illustration, what can be said about molecules of compounds and the kind of atoms they contain?

15. Of what does a water molecule consist?

16. What kind of molecule is made up of one hydrogen atom and one chlorine atom?

EXERCISE 5 ▮▮▮▮▮▮▮▮▮▮▮▮▮▮▮▮▮▮▮▮▮▮▮

Use Table 5-1 on page 248 to answer Questions 17-20.

17. What is the title of this table?

18. What are the horizontal rows called?

19. What are the vertical columns called?

20. What information is included in each of the squares in the table?

EXERCISE 6 ▮▮▮▮▮▮▮▮▮▮▮▮▮▮▮▮▮▮▮▮▮▮▮

Use Figure 5-2 on page 249 in this book to answer Questions 21-25.

21. What is the title of this graph?

22. What do the numbers on the vertical scale stand for?

23. What do the numbers on the horizontal scale stand for?

24. What is the chemical symbol for the alkali metal that has an atomic number of approximately 37?

25. What is the melting point of Na?

EXERCISE 7 ▰▰▰▰▰▰▰▰▰▰▰▰▰▰▰▰▰▰▰▰▰

Use Figure 7-1 on page 252 to answer Questions 26-29.

26. What is the title of this diagram?

27. What connects the mercury reservoir to the gas buret?

28. What is the purpose of the center rod?

29. What units are used to measure the volume of gas in the gas buret?

EXERCISE 8 ▰▰▰▰▰▰▰▰▰▰▰▰▰▰▰▰▰▰▰▰▰

Use Table 7-1 on page 253 in this book to answer Questions 30-35.

30. What is the title of this table?

31. Define Torr.

32. What information is provided in the columns?

33. If the pressure is 750 torr (at constant temperature) what will the volume be?

34. What can be said about the numbers in the column labeled PV Product?

35. As the pressure increases, what happens to the volume?

EXERCISE 9 ■■■■■■■■■■■■■■■■■■■■

Use Figure 7-2 on page 253 in this book to answer Questions 36-40.

36. What is the title of this graph?

37. What do the numbers in the horizontal scale indicate?

38. What do the numbers on the vertical scale indicate?

39. As the pressure decreases, what happens to the volume?

40. When the volume is approximately 50 ml, what is the approximate pressure?

SUMMARY ■■■■■■■■■■■■■■■■■■■■

1. Most textbooks include figures and tables. A figure or a table can help you understand the written ideas more clearly. It can also help you remember them more easily.

2. A figure or table is usually very closely related to the written material in a chapter. It's important to look at the figures and tables as you read. Try to see how each figure or table is related to the written text.

3. In most textbooks, the figures and tables are numbered and titled. The title tells you what the figure or table is about. The title usually appears in large print. It may be at the top, at the bottom, or at the side of a figure or table.

4. The source of a figure or table may be given. The source tells where the information came from and its date.

5. A diagram is a drawing, or sketch. There are many different kinds of diagrams. Some diagrams show an object; others show the steps in a process. When you study a diagram:
    a. Read the title to see what the diagram shows.
    b. Usually, there is a written explanation of the diagram. Examine the parts of the diagram as you read the explanation.

6. A graph presents numerical or statistical information in picture form. Usually, graphs are used to help you compare two or more quantities.

7. A circle graph is often called a "pie graph." It is used to show how a quantity is divided into portions. When you study a circle graph:
    a. Read the title and explanation to see what the graph is about.
    b. Make sure you understand what quantity the whole circle stands for. Then read the labels to see how the circle is divided into portions.

8. A bar graph uses a series of bars to help you compare several quantities or amounts. The bars may be either vertical or horizontal. When you study a bar graph:
    a. Read the title and explanation to see what the graph is about.
    b. Look at the vertical and horizontal scales to see what the bars represent. One scale will identify the bars; the other will be like a ruler, used to measure the length of the bars.
    c. Use the scales to see what quantity each bar stands for. Then compare the bars to one another.

9. A line graph is most often used to show how some quantity increases or decreases over time. There can be one line on a line graph, to represent one quantity; or there can be several lines, to represent different quantities. When you study a line graph:
    a. Read the title and explanation, to see what the graph is about.
    b. Look at the vertical and horizontal scales to see what the line(s) represents.
    c. Use the scales to see what different points on the line stand for. If the graph has more than one line, compare the lines to one another.
    d. Look at the direction of the line – up usually represents increases, down represents decreases. Then look for a pattern of increases or decreases.

10. A table is used to present a large amount of detailed information in a compact, easy-to-read form. Tables usually show numerical or statistical information. When you study a table:
    a. Read the title and explanation to see what the table is about.
    b. Read the column headings to see what information is in each column. Then read across a row, looking at the information in each column. If there are notes at the bottom of the table, they can help you understand the entries in the table.
    c. In most tables, you can compare the information in various ways. Read across the row, and compare the information in one column to the information in the other columns. Or read down a column, and compare the different entries to one another.

6
Taking Notes from Reading Assignments

There are times when you may not want to mark your textbook. Taking notes from a reading assignment is a good way to identify and organize the information you need to learn.

APPLYING THIS SKILL

To practice applying this skill, you need:

1. This book
2. A pen
3. A notebook

Take these items to a work area. Read the review section to make sure you understand the information you need in order to apply this skill. Then complete the exercises for Skill 6.

PERFORMANCE OBJECTIVE

When you have completed Skill 6, you will be able to take notes from science reading assignments. You will be able to choose and write down the important information and organize your notes in outline form.

Taking Notes From Reading Assignments

There are times when you may not want to mark your textbook. Taking notes from a reading assignment is another way to identify and organize the information you need to learn. It's actually easier to remember the information in a set of notes which you've written because taking notes requires you to think through the material, condense it, and organize it on paper.

A sample outlined section from a science textbook is included at the end of this review. You may refer to it as you review the important points listed below for applying this skill.

Let's review now what you need to know in order to take notes from reading assignments:

1. A good system for taking notes is outline form. Outline form shows you how each piece of information is related to the other pieces of information. Outline form involves using a series of numbers and letters along with indenting. Numbers and letters clearly show the different levels of indented information allowing you to easily see how the information is related. When you take notes in outline form from a textbook chapter, the organization of your notes will follow the same basic pattern as the headings in the chapter. See the explanation below.

   The Chapter Title Should be Centered at the Top of the Page

   I. The main section headings should be written beside Roman numerals.
      A. Subsection headings, which usually have paragraphs right after them, should be placed beside capital letters. You will turn these headings into questions and read to find the answers. Section headings beside A, B, etc. are about I.
         1. Answers to the questions should be written beside one or more numbers in your notes. Information beside 1, 2, etc. is about A.
            a. Information about the answers should be placed beside small letters in your notes. This information is about 1.
               (1) Additional details about "a" may be placed beside numbers in parentheses.
                  (a) If you want to add more details about (1), you may write them beside small letters in parentheses.

2. When you take notes from chapters in outline form, it's helpful to include information signaled by:

   a. headings
   b. key terms and definitions
   c. lists
   d. illustrations

3. Be sure to include in your outline any key terms and their definitions. Place them at one of the levels in the outline. Where they fit in the outline depends on how the information is organized in the chapter. Write the term; place DEF after it in the outline to call attention to it; then write its definition.

4. When information is provided in lists, include it in your notes. Place the words that tell you what a list is about at one of the levels in the outline. Where they fit depends on how the information is organized in the chapter. Under these words write the items listed.

5. You may want to include in your outline information from some illustrations. You might sketch the illustrations in your notes if that's possible. You also might xerox them and staple them to your notes. If you own the textbook or material from which you're taking notes, it is also helpful to simply write "See Illus. pg. ___" beside the appropriate line(s) in your notes. Then you can quickly refer to the illustration, a map, graph, diagram or table, in your textbook when you study your notes later.

6. You may include in your notes other information in addition to that signaled by headings, key terms and definitions, lists and illustrations. Your instructor may point out information to look for in the assignment.

7. In outlined notes, keep the wording short but clear. Abbreviate words and use symbols whenever possible.

   As when marking, be prepared to include numerous terms and their definitions and many facts and dates within your notes from science textbooks.

Below is a sample outline of a section from a physics textbook:

Classification

I. Electromagnetic Waves
    A. Electromagnetic waves in general
        1. electromagnetic waves - DEF - electric and magnetic
            fields which travel outward and behave like waves
        2. 2 points about waves
            a. each has alternating electric and magnetic fields at
                right angles to each other
            b. frequency varies
                (1) frequency - DEF - # of oscillations per sec
    B. Specific electromagnetic waves
        1. light - differs from others due to its frequency/wavelength;
        2. 5 other electromagnetic waves - some longer, some shorter
            than visible light
            a. ultraviolet light (UV) - just shorter
            b. infrared (IR) - just longer
            c. radio waves - much longer
            d. x-rays - very long (extremely high frequency)
            e. gamma rays - longer yet (even higher frequency)

See Illus.   C. Electromagnetic spectrum
pg. ___   1. electromagnetic spectrum - DEF - classification of
                       electromagnetic waves according to wavelength/frequency

EXERCISES

There are 9 exercises for Skill 6. In each exercise, you will take notes from a reading assignment in one of
the sample chapters.

In Exercises 1-3, you'll read sections from chapters on biology. Exercises 4-6 deal with physics. Exercises
7-9 are about chemistry. You may complete as many exercises as you need, and you may work on the
exercises in any order.

The directions for each exercise tell you which sample chapter to use, and which section to read. Be sure to
find the correct textbook section before you begin taking notes. Write your notes in outline form in your
notebook. Start each exercise on a new page. You can compare your outlines with those in the answer key
at the back of this book. When comparing outlines, remember that the levels of indenting are more important than the exact numbering and lettering of your outlines. You should include all important information
in your notes. You may choose not to include all of the examples shown in the outlines in the answer key.

For these exercises it is unnecessary to write the chapter titles in your notes. In each of these exercises, only
one section from a chapter has been included.

EXERCISE 1 ▟

You will use part of the biology chapter that begins on page 215 in this book. The chapter title is "Algae."

1. Locate the first major heading in the chapter, What are Algae Like?, on page 215.
2. Begin taking notes at this major heading. The first line in your notes will be:
   I. What are Algae Like?
3. Continue taking notes until you reach the second major heading, Blue-green Algae, on page 215.
4. Check your notes with the answer key for Skill 6 in the back of this book.

EXERCISE 2 ▟

You will use part of the biology chapter that begins on page 219 in this book. The chapter title is "Protozoa."

1. Locate the third major heading in the chapter, Ciliates, on page 221.
2. Begin taking notes at this major heading. The first line in your notes will be:
   III. Ciliates.
3. Continue taking notes until you reach the chapter summary on page 222.
4. Check your notes with the answer key for Skill 6 in the back of this book.

EXERCISE 3 ▟

You will use part of the biology chapter that begins on page 223 in this book. The chapter title is "Origin of Genetics."

1. Locate the second major heading in the chapter, Further Experimentation, on page 225.
2. Begin taking notes at this major heading. The first line in your notes will be:
   II. Further Experimentation.
3. Continue taking notes until you reach the chapter summary on page 225.
4. Check your notes with the answer key for Skill 6 in the back of this book.

EXERCISE 4 ■■■■■■■■■■■■■■■■■■■■■■■■■■■■■■

You will use part of the physics chapter that begins on page 227 in this book. The chapter title is "Speed, Velocity, and Acceleration."

1. Locate the first major heading in the chapter, Speed, on page 227.
2. Begin taking notes at this major heading. The first line in your notes will be:
    I. Speed.
3. Continue taking notes until you reach the second major heading, Velocity, on page 228.
4. Check your notes with the answer key for Skill 6 in the back of this book.

EXERCISE 5 ■■■■■■■■■■■■■■■■■■■■■■■■■■■■■■

You will use part of the physics chapter that begins on page 231 in this book. The chapter title is "Newton's Three Laws of Motion."

1. Locate the first major heading in the chapter, Newton's First Law of Motion, on page 231.
2. Begin taking notes at this major heading. The first line in your notes will be:
    I. Newton's First Law of Motion.
3. Continue taking notes until you reach the second major heading, Newton's Second Law of Motion, on page 231.
4. Check your notes with the answer key for Skill 6 in the back of this book.

EXERCISE 6 ■■■■■■■■■■■■■■■■■■■■■■■■■■■■■■

You will use part of the physics chapter that begins on page 237 in this book. The chapter title is "Work."

1. Locate the second major heading in the chapter, Work Done Aginst Gravity, on page 239.
2. Begin taking notes at this major heading. The first line in your notes will be:
    II. Work Done Against Gravity.
3. Continue taking notes until you reach the chapter summary on page 241.
4. Check your notes with the answer key for Skill 6 in the back of this book.

EXERCISE 7 ▰▰▰▰▰▰▰▰▰▰▰▰▰▰▰▰▰▰

You will use part of the chemistry chapter that begins on page 243 in this book. The chapter title is "Atoms and Molecules."

1. Locate the second major heading in the chapter, Molecules, on page 244.
2. Begin taking notes at this major heading. The first line in your notes will be:
   II. Molecules.
3. Continue taking notes until you reach the chapter summary on page 245.
4. Check your notes with the answer key for Skill 6 in the back of this book.

EXERCISE 8 ▰▰▰▰▰▰▰▰▰▰▰▰▰▰▰▰▰▰

You will use part of the chemistry chapter that begins on page 247 in this book. The chapter title is "Periodic Table of Elements."

1. Locate the first major heading in the chapter, Rows, on page 247.
2. Begin taking notes at this major heading. The first line in your notes will be:
   I. Rows.
3. Continue taking notes until you reach the second major heading, Groups, on page 247.
4. Check your notes with the answer key for Skill 6 in the back of this book.

EXERCISE 9 ▰▰▰▰▰▰▰▰▰▰▰▰▰▰▰▰▰▰

You will use part of the chemistry chapter that begins on page 251 in this book. The chapter title is "Relationship of Volume and Pressure of Gases."

1. Locate the second major heading in the chapter, Conclusions, on page 254.
2. Begin taking notes at this major heading. The first line in your notes will be:
   II. Conclusions.
3. Continue taking notes until you reach the third major heading, Common Sense, on page 255.
4. Check your notes with the answer key for Skill 6 in the back of this book.

SUMMARY

1. When you take notes from a reading assignment, choose and write down only the important information. Organize it in a way that will help you learn and remember the material.

2. Be sure to include the important information signaled by headings, key terms and definitions, lists, and illustrations. You should also include other information that you think is important. Sometimes you may want to include examples or explanations that help make important ideas clear. You may also want to sketch an illustration in your notes or write "See Illus. pg.___" in the margin.

3. A good system for organizing notes is outline form. Outline form shows you what the important information is and how each item of information is related to the other items of information about a topic. While the numbering and lettering of information in an outline may vary greatly, the levels of indenting should show accurately the relationship of information.

4. When you take notes from a reading assignment that has headings, use the headings to focus your attention on important information and to organize your notes.

7
Taking Notes from Lectures

While taking science courses, you will listen to many lectures about various topics. You have to listen actively for the information you need to learn, and you have to organize your notes well for study.

APPLYING THIS SKILL

To practice applying this skill, you need:
1. This book
2. A pen
3. A notebook
4. The audiocassette labeled "The Sciences, Lectures for Exercises."

Take these items to a work area with a tape player. Read the review section to make sure you understand the information you need in order to apply this skill. Then complete the exercises for Skill 7.

PERFORMANCE OBJECTIVE

When you have completed Skill 7, you will be able to take organized notes on important information in science lectures.

Taking Notes From Lectures   ▰▰▰▰▰▰

While taking courses in the sciences, you will listen to many lectures about various topics. To listen actively, it's helpful if you've read the assigned material before class. Then during the lecture, it's helpful to think about what the lecturer says and how it relates to the reading assignment. At the same time, it's important to take notes on the important information. A sample outlined lecture is included at the end of this review. You may refer to it as you review the important points listed below for applying this skill.

Let's review now what you need to know in order to take good notes from lectures.

1. When you listen to lectures, you should listen and watch for several signals of important information.
   a. The first signal of important information is an introduction to the topic of the lecture. For example, a lecturer may often introduce the topic by saying something like, "Today, I'm going to discuss..." The topic is then mentioned.
   b. Another signal of important information will be the words introducing statements of importance. A lecturer might say things like:

   "An important feature..."
   "The chief reason..."
   "You should remember that..."
   "It's important to know that..."
   Statements of importance will follow these words.
   c. A third signal of important information would be the words signaling definitions of terms. Before giving actual definitions a lecturer may say something like:

   "The definition is..."
   "...is called..."
   "...is referred to as..."
   d. A fourth signal of important information would be words that introduce lists, telling you what the items in the lists are about. Lecturers use words like:

   "There are many reasons..."
   "The five steps are as follows..."
   "There are a number of ways..."
   to introduce lists. Then, each item in a list may also be signaled by words such as:

   "First — Second — Third..."
   "First — Next — Then — Finally..."
   "One — Another — And another..."

e. A fifth signal of important information would be changes in speed and volume. A lecturer may slow down and speak louder or drop his or her voice level to stress a point.

f. A sixth signal of important information would be the words signaling summaries or restatements of important points. At the end of lectures, a lecturer may say something like:
"To summarize..."
"Let's review what we've covered today..."

g. A final signal of important information is one you need to look for while you listen to the lecture. Lecturers may use visual aids such as transparencies and maps or charts. They may make sketches and write terms on the board also. Information in visual aids and items placed on the board usually should be included in your notes.

2. As you listen and watch for these signals, you should also follow a procedure when taking notes from lectures.

a. First, put the date of the lecture in the upper right-hand corner of your notes.

b. Second, write the topic of the lecture as the title of your notes. Write it in the center at the top of the page.

c. Third, write important points that are main ideas next to the left-hand margin. The signals of important information listed above may alert you to main points or ideas.

d. Fourth, under the main points indent the major details that explain or support them. The same signals of important information are used to alert you to major details. Whether the statement is a main point or a detail depends on how the information is organized in the lecture.

e. Fifth, under major details, you may need to indent additional details that explain or support them. Details at this level are often items in a list.

f. Terms that are defined in a lecture should always be included in your notes. Whether they are main points or details depends on how the information in the lecture is organized. When you include a term, write the term first; write DEF after it; then write its definition.

g. You may also include examples or explanations in your notes, but often it is not necessary to do so.

    h. Remember to include somewhere in your notes the information
        in sketches or illustrations if they are not too detailed.
        Write labels for your illustrations.

    i. Finally, when you hear a lecturer summarizing or reviewing
        important information, go back to the beginning of your notes as
        you hear the summary to see if you've included all of the
        information. Add any points which you may have missed.

3. When following this procedure for taking notes, remember these four points:

    a. Be brief. Write the key words of the sentences you hear.
        Abbreviate words and use symbols when possible.

    b. If you miss something, continue taking notes and fill in the gaps
        later.

    c. If you have trouble spelling a word, continue taking notes and
        correct the spelling later.

    d. Try to write neatly.

4. After a lecture, review your notes as soon as possible in order to:

    a. Add information you didn't have time to write down.

    b. Discover if something in your notes is unclear to you.

    c. Organize your notes in outline form by adding Roman
        numerals, letters, and numbers to your notes to make
        the relationship between main points and details as
        clear to you as they are in your notes from written
        material. When you do this:

        I. Use Roman numerals (I, II, etc.) for main ideas
           A. Use capital letters (A, B, etc.) for major details about I,
              II, etc.
              1. Use regular numbers (1, 2, etc.) for additional details
                 about A, B, etc.
                 a. Use small letters (a, b, etc.) for any details
                     about 1, 2, etc.
                    (1) Use numbers in parentheses to add further
                         information about "a".
                      (a) If you want to add more details about (1),
                           (2),etc., use small letters in parentheses.

Below is a sample outline of a lecture in chemistry.

A Classification System in Chemistry

I. A Major Goal of Chemists: Classify or Arrange Elements
II. History of Periodic Table
    A. Mendeleyev - Russian Chemist, 1843-1907
        1. established arrangement of elements called Periodic Table
            a. allowed spaces for undiscovered elements - eg.
            b. predicted properties for some unknown elements like
                atomic weights - eg.
    B. Modern Periodic Table - classifies all elements by placing
        them in groups having similar properties
        1. 2 points about modern table
            a. elements arranged in consecutive order; consecutive
                number of element used to identify it
                (1) atomic # - DEF - consecutive order # of element
                      in table starting with hydrogen, lighest element
            b. similar properties recur periodically, so elements
                can be grouped in columns
                (1) natural families - DEF - groupings of elements
                      in vertical columns - basis of Periodic Table

EXERCISES ▬▬▬▬▬▬▬▬

There are 7 exercises for Skill 7. For these exercises, you will need the audiocassette labeled "The Sciences, Lectures for Exercises," a pen, and your notebook.

The lectures for Exercises 1-3 are on topics from biology. The lectures for Exercises 4-6 deal with physics. The lecture for Exercise 7 is about chemistry. You may complete as many exercises as you need, and you may work on the exercises in any order.

Start each exercise on a new page in your notebook. Listen to the lecture and take notes. At the end of each lecture, stop the tape and review your notes. Add numbers and letters to your notes to organize them in outline form.

You can compare your outlines with those in the answer key at the back of this book. Play the tape again as you look at the answer key and your own notes. When comparing outlines, remember that showing the relationship of information is more important than the exact numbering and lettering of your outlines. You should include all important information in your notes. You may choose not to include all of the examples shown in the outlines in the answer key.

SUMMARY

1. In order to be an active listener at lectures, you should:
    a. Prepare yourself for a lecture by reading assigned material ahead of time.
    b. Think about how the previous lecture relates to the current one and to your reading assignments.
    c. Focus your attention on the lecturer.
    d. Take notes on the important information in the lecture.

2. During a lecture, you should listen and look for these signals of important information:
    a. introduction
    b. statements of importance
    c. definitions
    d. lists
    e. changes in speed and volume
    f. summary
    g. visual aids

3. When you take notes from a lecture, keep these guidelines in mind:
    a. Be brief. Use abbreviations whenever they are helpful.
    b. If you miss some information, continue taking notes and fill in the gaps later.
    c. Try to write neatly.
    d. Show how the information in your notes is related by putting main points next to the margin and indenting the details.

4. After a lecture, review your notes as soon as possible. Add any important information that you did not have time to write down. Reword or check any information in your notes that is not clear to you. Adjust your notes as needed to organize them in outline form.

Additional Suggestions for Science Students

Science instructors often provide demonstrations during class lectures. For example, they may bring actual materials to the classroom to demonstrate scientific principles. To fully understand class lectures and demonstrations, it's important to read any material assigned before coming to class and to complete the related section in the study guide if one accompanies your textbook.

Instructors often ask questions during classroom lectures. They may, for example, ask students to explain how scientific principles are applied. Class participation is expected.

Lab work is usually required in addition to classroom lectures. Often experiments are assigned to confirm textbook and lecture theory. Instructors, however, may provide short lectures in the lab before experimental work begins to review what students need to know. In addition, they may demonstrate procedures and use of equipment so students know what they're going to do and how to do it. To understand and describe results of these experiments, it's important to know the scientific principles involved. It's most helpful to read assigned material before classroom lectures and related lab sessions.

On the other hand, some experiments are assigned to introduce scientific concepts. In other words, a discovery approach is used and students must carefully observe reactions and note results to arrive at the principle involved. Later, lecture and textbook information is provided to help students clarify their analyses.

8

Taking Tests

When test time draws near, it is important to prepare well in order to show that you understand and remember what you've been studying. Many tests in science courses contain both objective questions and essay questions. You need to know the techniques that can help you take both kinds of tests effectively.

APPLYING THIS SKILL

To practice applying this skill, you need:

1. This book
2. A pen or pencil

Take these items to a work area. Read the review section to make sure you understand the information you need in order to apply this skill. Then complete the exercises for Skill 8.

PERFORMANCE OBJECTIVE

When you have completed Skill 8, you will be able to apply specific techniques for taking objective tests and essay tests in science courses.

Taking Tests

When test time draws near, it is important to prepare yourself to take tests in order to better show that you understand and remember what you've been studying. In general, there's a sequence of steps to follow when preparing for tests. First you should find out as much as you can about the test. Always complete this step at least a week ahead of time. Then you should gather all the material that you will need to study and review it, noting important information. Next you should combine your material into an outline. Recall and recite the material in the outline daily. Reciting or saying the information aloud helps you understand and remember what you need to know.

Many tests for which you must prepare will contain both objective questions and essay questions. Objective questions measure your knowledge of specific information. There is only one correct answer. You need to memorize facts and details for those questions. Quite often science students are given objective questions. Some problems to solve may be included too, particularly on physics and chemistry tests. Sometimes, science students are given short essay questions as well. Essay questions measure your knowledge of a particular topic. There may be several ways the question can be answered, so there isn't just one correct answer. Your instructor, however, will look for certain information to be included in an answer, organized around a main topic. To prepare for essay tests you need to study the main topics and see how facts and details under them relate to one another. When you actually take tests then, you either indicate your knowledge of specific information, or you provide in your own writing a detailed answer on a particular topic. Both objective and essay questions are given to science students; however, more objective questions can be expected than essay questions.

It's most important too, when taking tests, to read both the directions and questions very carefully for objective and essay items.

In general, follow these guidelines when taking tests:

1. Read all test directions carefully and survey the entire test before answering any questions.
2. Budget your time so you will be able to complete the entire test.
3. Read the questions carefully and answer those you're sure of first. If there's no penalty for wrong answers, guess.
4. Try to save time to review your answers before submitting your test.

Now let's review what you need to know in order to answer objective questions and essay items.

OBJECTIVE QUESTIONS ▮▮▮▮▮▮▮▮▮▮▮▮▮▮▮▮▮

1. When you answer objective test questions, you should answer those you are sure of first. Then, if there is no penalty for wrong answers, guess.

2. The most common kinds of objective test questions are true-false, multiple-choice, matching, and short-answer.

3. To answer a true-false question, you must decide if a statement is true or false. When a true-false statement contains an absolute, the statement is usually false. Absolutes are words that say there are no exceptions, such as *always, never, all*, and *none*. When a true-false statement contains a qualifier like *generally* or *often*, the statement is usually true. To simplify a true-false question that contains a double negative, try crossing out both negatives. Then decide if the statement is true or false. If any part of a true-false question is false, the entire statement is false.

4. In a multiple-choice question, you are given a number of possible answers. You must choose the best answer from the choices given. The directions should tell you whether or not you may include one or more of the choices given. If your answer may include more than one choice, the directions given may read something like: "Write the letter or letters of the answer in the blank." When you guess the answer to a multiple-choice question, use the process of elimination. To do this, you:

    (1) Read all the answer choices.
    (2) Eliminate, or drop, the choices that you know are wrong. You can usually eliminate answers that are too general or that are only partially right.
    (3) Guess the answer from the remaining choices. When your answer must be only one of the choices in a multiple-choice question, "all of the above" is the answer to choose if you're sure that at least two other choices are correct. "All of the above" is always the wrong answer for a multiple-choice question if you're sure that one other choice is wrong.

    In a multiple-choice question, "All of the above" is a good guess if your instructor does not give this choice regularly. Another clue is the length of the choices. The longest choice is often the correct answer.

5. In a matching test, you must match each item in one list with an item from another list. The directions will tell you if an item can be used more than once.

6. In a short-answer question, you have to supply the answer on your own. There are no choices given. If you don't know the answer to a short-answer question, you may be able to guess the answer from clues in the question. Sometimes, the number of lines is a clue to the number of words in the answer. The word *an* before a blank line tells you that the answer begins with a vowel sound.

ESSAY QUESTIONS

1. Four steps that will help you budget your time when you take a test including essay questions are:
   a. Read all the questions first.
   b. Use a system of check marks to decide which questions to answer first. Put 3 check marks by the questions that you know you can answer and that have a high point value. Answer these questions first. Put 2 marks by the questions that you know you can answer, but that don't have such a high point value. Answer these questions next. Then, try to answer those questions that have a high point value, even though you may not be able to answer them easily. You might have put 1 check mark beside these questions. Finally, answer the remaining questions for which you didn't indicate any check mark.
   c. Try to outline the answers to questions that you don't have time to answer.
   d. If you have time at the end of the test, read your answers and see if you need to add or change anything.

2. To make an essay answer clear, organize your thoughts before you write. Make a list of the important points or write an outline for the answer before you respond. Turn the essay question into a statement and use the statement to begin your answer. Write neatly and follow your outline.

3. Six direction words to look for in essay questions on science tests are:
   a. define: the correct answer must include the meaning of a term.
   b. compare: the correct answer must show the similarities and differences.
   c. contrast: the correct answer must show the differences between two or more things.
   d. explain: the correct answer must give the reasons for the causes of a certain thing.
   e. discuss: the correct answer must include information about the subject from some point of view(s).
   f. describe: the correct answer must include details that will allow the reader to form a picture of the subject.

   Each of these words requires a special kind of answer.

4. Leave space at the end of each answer, so you may add more information later if necessary.

5. Save some time to review your answers.

EXERCISES ▰▰▰▰▰▰▰▰▰▰▰▰▰▰

There are 13 exercises for Skill 8. You will answer objective test questions, and you will judge the answers to several essay questions. All of the exercises are based on sections from the sample chapters. In each exercise including objective questions, review the reprinted section before you answer the questions. Write your answers in the spaces provided in your book. You can check your answers with the answer key at the back of the book.

EXERCISE 1 ▰▰▰▰▰▰▰▰▰▰▰▰

Questions 1-2 are based on sections from the sample chapter "Algae." Review the chapter section which is reprinted below. Then answer the test questions.

> *Four general traits.* All algae share four traits. First, algae possess chlorophyll and carry on photosynthesis. *Chlorophyll* can be defined as the green pigment needed for metabolic processes in plants. *Photosynthesis* is the process by which light energy is absorbed and then converted to the chemical bond energy of glucose. Second, algae can live in both fresh and salt water. Third, algae are, in general, simple organisms lacking both specialized tissues for conducting water and plant organs such as roots, stems, and leaves. Fourth, they have complex reproductive processes. They certainly must be considered successful forms of life.

Choose the best answer for each question and circle the letter of your choice.

1. Algae
   a. possess chlorophyll and carry on photosynthesis.
   b. can live in both fresh and salt water
   c. lack specialized tissues for conducting water and plant organs such as roots, stems, and leaves
   d. carry on complex reproductive processes
   e. all of the above

2. Chlorophyll can be defined as
   a. green pigment
   b. green pigment needed for food manufacture in plants
   c. green pigment needed for food manufacture in ferns

EXERCISE 2 ▆▆▆▆▆▆▆▆▆▆▆▆▆▆▆▆▆▆▆▆▆▆

Questions 3-5 are based on a section from the sample chapter "Protozoa." Review the chapter section which is reprinted below. Then answer the test questions.

Protozoa in General

All protozoa share certain characteristics. First, they are one-celled, nongreen, animal-like protists. Second, they can move about and are classified according to how they move. Third, protozoa must secure their own food and are specially equipped to do so. Most protozoa actively trap their food. They get it from the surrounding water and consume it in chunks. Some protozoa, however, are parasites. They live directly off other organisms.

Amoebae

Forming pseudopodia. One of the simplest protozoans is the amoeba. Amoebae are unique in how they move. To move, they develop pseudopodia. *Pseudopodi*a are false feet formed by flexible plasma membranes and cytoplasm. The formation of pseudopodia involves certain stages. Initially, the cytoplasm is in a fixed, gelatinous state. Then, chemical changes cause part of the cytoplasm to become a fluid. This fluid, which is like a true solution, is known as a sol. Cytoplasm in a sol condition can flow freely. Next, cytoplasm flows into the cell membrane, extending it to form a pseudopodium. Then this cytoplasm changes back to a gel. Finally, other areas of the cytoplasm undergo the same sol-gel transformation forming new pseudopodia. New pseudopodia are formed and others disappear. In this way, the shape of amoebae changes constantly.

Circle T if a statement is true or F if a statement is false.

3.   T or F   All protozoa develop pseudopodia to move about.

4.   T or F   All protozoa are one-celled, nongreen, animal-like protists.

5.   T or F   All protozoa are parasites.

EXERCISE 3 ▰▰▰▰▰▰▰▰▰▰▰▰▰▰▰▰▰▰▰▰▰▰▰

Question 6 is based on a section from the sample chapter "Speed, Velocity, and Acceleration." Review the chapter section which is reprinted below. Then answer the question.

Speed

Speed and its formula. Speed is the most commonly used rate. The amount of *speed* is a distance traveled in some period of time. The following formula can be used:

$$speed = \frac{distance\ traveled}{time\ period} \ or\ s = \frac{d}{t}$$

Consider an example. A runner completes a 100-yard dash in 10 seconds. His speed would be:

$$s = \frac{100\ yd}{10\ sec} = \frac{10\ yd}{sec}$$

Each automobile has a speedometer. It continuously shows the speed of the auto. An American automobile's speedometer shows the speed in miles/hour.

Distance. Knowing what the speedometer reads allows you to figure the distance a car will travel in a certain time period. Consider an example. Suppose the speedometer reads 60 miles/hour, and this speed stays the same. In a time period of three hours, the car will travel a distance of 180 miles. The formula used to figure distance is:

$$distance = speed \ x \ time \ or \ \mathbf{d = st}$$

In the example stated above:

$$\mathbf{d} = 60\ mi \ x \ 3 \ hrs \ or \ 180 \ miles$$

Velocity

The concept of speed avoids direction. Many situations, however, require one to know the direction in which an object travels. It is just as important as knowing the distance. For example, a pilot may need to fly from New York to Chicago. He'll want to know both the speed and direction of the plane. The *velocity* of a body is a specification of both its speed and the direction in which it is moving. A car, for example, may travel due north at 50 miles/ hour. Another car may travel due east at 50 miles/hour. They would travel at the same speed. Their velocities, however, would be different. When direction is implied, velocity is the term to use.

Acceleration

Acceleration response. Every driver knows that he must press down the accelerator pedal to increase the speed of his car. All cars do not, however, respond in the same way to a depression of an accelerator. Some will change speed much faster than others. The *acceleration response* refers to the time it takes to achieve a certain speed starting from rest. If you compare the acceleration of two cars, the one reaching a specified speed in the least time has the greater acceleration. Acceleration can be expressed by the formula:

$$\text{acceleration} = \frac{\text{change in speed}}{\text{time required for change to occur}}$$

Direction and acceleration. Direction can be part of acceleration. Acceleration then refers to a change in velocity rather than speed. You can say then that an object can be accelerated by changing either (or both) the speed and direction.

6.   Choose one description from Column B to match each term in Column A. Write the letter from Column B beside the correct number of its match in Column A. You may not use a letter more than once. Some items in Column B will not be used.

Column A   Column B

_____1. speed   a. speed x time
_____2. velocity   b. a specification of a body's speed and the
_____3. acceleration   direction in which it is moving
_____4. distance   c. a force between the objects, each having the
                               physical properties of mass
                       d. the rate at which distance is being covered by
                               a body
                       e. the change in some quantity divided by the time
                               required to produce the change
                       f. the rate at which a body's speed or velocity
                               changes with time

EXERCISE 4 ██

Questions 7-9 are based on a section from the sample chapter "Speed, Velocity, and Acceleration." The section is reprinted at the start of Exercise 3. Review the section. Then answer the test questions, writing the correct word(s) in each blank.

7.   The amount of _____ is a distance traveled in some period of time.

8.   The formula used to figure distance is _____.

9.   Velocity specifies two things about a body. What are they?
     _____ and _____.

EXERCISE 5 ██

The essay question below is based on the sample chapter "Algae." If you want to, you can turn to page 215 and review the sample chapter. Next, read the essay question and the answer. Then answer Questions 10-12 by writing Yes or No. Answer Question 13 if you want to practice writing essay answers.

Question:
Describe two things about blue-green algae: their cell structure and how they reproduce.

Answer:
There isn't a definite nucleus in a single cell. Therefore, DNA is scattered throughout the cell. There are no chloroplasts to contain chlorophyll so the chlorophyll is attached to membranes. Also, single cells are arranged in filaments or chains. There is very little division of labor among the cells. Actually, each cell in the chain is like every other cell in that group. Finally, groups of cells are often enclosed within a protective jelly-like layer.

There is no sexual reproduction with blue-green algae. Instead, they reproduce asexually through division of a single cell. New cells are formed along a chain or filament through simple fission. Fission is reproduction in which a one-celled organism divides into two one-celled organisms. Sometimes, filaments break apart. Each section then produces a new series of cells through fission. Finally, there's one kind of blue-green algae which produces spores, specialized cells. These spores can develop into new filaments.

10. Has the essay question been turned into a statement to begin the answer?
     If no, write a sentence that would be a good way to begin the answer.

11. Is the answer the kind required by the direction word in the question?

12. Does the answer include all the points required by the question?

13. (Optional) To practice taking essay tests, write your own answer to this question. Use your
     own paper. Then ask your instructor to judge your answer.

EXERCISE 6 ▮▮▮▮▮▮▮▮▮▮▮▮▮▮▮▮▮▮

The essay question below is based on the sample chapter "Protozoa." If you want to, you can turn to page
219 and review the sample chapter. Next, read the essay question and the answer. Then answer Questions
14-16 by writing Yes or No. Answer Question 17 if you want to practice writing essay answers.

Question:
Define "pseudopodia" and explain two reasons why they're important to amoebae.

Answer:
Pseudopodia can be defined as false feet formed by flexible plasma membranes and cytoplasm. Pseudopodia are important to amoebae for two reasons. First, they enable amoebae to
move about. Initially, the cytoplasm is in a fixed, gelatinous state. Chemical changes then
cause part of the cytoplasm to become a fluid, known as sol, which is like a true solution. Cytoplasm in a sol condition can flow freely. It flows into the cell membrane extending it to
form a pseudopodium. Then the cytoplasm changes back to sol. Eventually, other areas of
the cytoplasm undergo the same sol-gel transformation forming new pseudopodia as others
disappear. The shape changes constantly as amoebae move about.

Second, pseudopodia are used by amoebae to get food. Pseudopodia engulf particles of food.
The food enters the cytoplasm where it's digested. The digestion actually takes place in the
vacuoles in the cytoplasm.

14. Has the essay question been turned into a statement to begin the answer?
     If no, write a sentence that would be a good way to begin the answer.

15. Is the answer the kind required by both direction words in the question?

16.   Does the answer include all the points required by the question?

17.   (Optional) To practice taking essay tests, write your own answer to this question. Use your own paper. Then ask your instructor to judge your answer.

EXERCISE 7 ▇▇▇▇▇▇▇▇▇▇▇▇▇▇▇▇▇▇▇▇▇▇▇▇▇▇▇▇▇▇▇▇

Questions 18-19 are based on a section from the sample chapter "Newton's Three Laws of Motion." Review the chapter section, which is reprinted below. Then answer the test questions.

> Remember as you read this chapter that the metric system of measurement is used in physics. The unit of length is the *meter*. One meter is equal to 3.28 feet. The *kilogram* is the unit of mass. One kilogram weighs 2.21 pounds at the earth's surface. Time is measured in *seconds*. This system is called the meter-kilogram-second, or MKS system.
>
> **Newton's First Law of Motion**
>
> *First law.* Newton's first law reads: "Every body continues in its state of rest, or of uniform motion, in a straight line unless it is compelled to change that state by forces impressed on it." *Uniform* here means constant. *Motion* refers to velocity. If the velocity is constant, both in magnitude and direction, the acceleration is zero. The velocity can be changed only by applying a force.
>
> *What this law suggests.* When an object is seen to be accelerating, Newton's first law suggests looking for the force that causes the acceleration. Often this force is obvious. At other times it is not. A batted baseball is accelerated by the force that the bat exerts on the ball. The force is obvious. A ball dropped from the window accelerates toward the ground. What is the force that accelerates it? The force is not so obvious. The answer, however, is known to be gravity. The gravitational force between the earth and the ball accelerates it.

Choose the best answer for each question and circle the letter of your choice:

18.   When considering acceleration and that which causes it, we should realize that:
      a.   acceleration of an object is great if the velocity is constant
      b.   acceleration of a batted baseball is caused by gravitational force
      c.   acceleration of a basketball is caused by the basketball player
      d.   all of the above

19. A kilogram is:
    a. the unit of mass in the metric system that weighs 2.21 pounds at the earth's surface
    b. a unit in the metric system
    c. a unit that weighs 2.21 pounds

EXERCISE 8 ▰▰▰▰▰▰▰▰▰▰▰▰▰▰▰▰

Questions 20-22 are based on a section from the sample chapter "Work." Review the chapter section which is reprinted below. Then answer the test questions.

> *Positive work and negative work.* It is important to recognize that a force can act in one of two ways. First, it can act in the same direction as the mass is moving (positive work). Second, force can act opposite to the direction in which the mass is moving (negative work). In both cases work is being done.

Circle T if a statement is true or F if a statement is false.

20.   T or F   Force always acts in the same direction as the mass is moving.

21.   T or F   Force always acts opposite to the direction in which the mass is moving.

22.   T or F   Force can act in the same direction as the mass is moving, and it can act opposite to the direction in which the mass is moving.

EXERCISE 9 ██████████████████████████

Question 23 is based on a section from the sample chapter "Origin of Genetics." Review the section which is reprinted below. Then answer the test question.

> **Results.** Mendel found that in every case the parental cross (P) yielded offspring with round seeds only. The offspring of a parental cross are called the *first filial,* or F_1, *generation.* There were no wrinkle-seeded plants in the F_1 generation.
>
> **Law of dominance.** Mendel concluded that for each trait there is one form which "dominates" the other. A *dominant trait* is the trait which appears exclusively in the F_1 generation. For example, the trait for round seeds is the dominant trait. It "dominates" the trait for wrinkled seeds. The trait which disappears in the F_1 generation is called a *recessive trait.* In this case, the trait for wrinkled seeds is the recessive trait. Mendel generalized this result and formulated the law of dominance. The Law of Dominance states that one trait, the dominant trait, dominates or prevents the expression of the recessive trait.
>
> **Further Experimentation**
>
> **F_2 generation.** The lack of wrinkle-seeded plants in the F_1 generation impressed Mendel. Therefore, he experimented further. He then crossed members of the F_1 generation. Some F_1 round-seeded plants were crossed with other F_1 round-seeded plants. The offspring of the F_1 generation cross are called the *second filial,* or F_2, *generation.*

23. Match one description from Column B with each term in Column A. Write the letter from Column B beside the correct number of its match from Column A. You may not use a letter more than once. Some items in Column B will not be used.

Column A

_____1. dominant trait
_____2. F_1 generation
_____3. recessive trait
_____4. law of dominance
_____5. F_2 generation

Column B

a. a trait that disappears in the F_1 generation
b. the study of heredity
c. states that one trait, the dominant trait, dominates or prevents expression of the recessive trait
d. offspring of a parental cross
e. the parental cross
f. offspring of the F_1 generation cross
g. a trait that appears exclusively in the F_1 generation

EXERCISE 10 ▬▬▬▬▬▬▬▬▬▬▬▬▬▬▬▬▬▬▬▬▬▬▬

Questions 24-26 are based on a section from the sample chapter "Work." Review the section which is reprinted below. Then answer the test questions.

> **Measuring work.** Work can be measured in units of foot pounds and joules. The foot pound (ft. lb.) is the unit of work in the British system of units. One *foot pound* is equal to the work done by a force of one pound acting through a distance of one foot.
>
> For example, a man may push a refrigerator. The amount of work he does is equal to the magnitude or amount of force he applies times the distance he moves the refrigerator. He may push the refrigerator 10 feet with a force of 20 pounds. If he does this, he does 20 lb. x 10 ft. = 200 foot pounds of work.
>
> The joule (J) is the unit of work in a more recent system of units, the Systeme International (SI), the current version of the metric system. One *joule* is equal to the work done by a force of 1N (newton) acting through a distance of 1m(meter), i.e. 1J = 1N x m. A *newton* is a metric unit of force that is equivalent to a force of about 1/4 pound.

Write the correct word(s) in each blank.

24.   One foot pound is equal to the work done by a _____ of one pound through
      a _____ of one foot.

25.   One joule is equal to the work done by a force of 1 newton acting through a distance of one
      _____.

26.   What is the newton, a metric unit of force, equivalent to?_____

EXERCISE 11 ████████████████████████

The essay question below is based on the sample chapter "Work." If you want to, you can turn to page 237 and review the sample chapter. Next, read the essay question and the answer. Then answer Questions 27-29 by writing Yes or No. Answer Question 30 if you want to practice writing essay answers.

Question:
Compare positive work and negative work and describe two situations, one illustrating positive work and one illustrating negative work.

Answer:
Positive work and negative work are similar in one respect, yet quite different. They are similar in that in both cases work is being done. They are, however, quite different. When positive work is being done, force is acting in the same direction as the mass is moving. When negative work is being done, force is acting opposite to the direction in which the mass is moving.

Two situations can be described, one illustrating positive work and one illustrating negative work. First, a force can act in the same direction as the car is moving. A car could rear-end another car slowly approaching a stop light. This would likely increase that car's speed; in fact, it would most likely end up in the intersection or beyond. Second, a force can act in a direction opposite to that in which the car moves; a car could hit another head-on. This would most likely decrease its speed. In fact, the car hit would most likely come to a stop.

27. Has the essay question been turned into a statement to begin the answer?
    If no, write a sentence that would be a good way to begin the answer.

28. Is the answer the kind required by both direction words in the question?

29. Does the answer include all the points required by the question?

30. (Optional) To practice taking essay tests, write your own answer to this question. Use your own paper. Then ask your instructor to judge your answer.

EXERCISE 12

The essay question below is based on the sample chapter "Speed, Velocity, and Acceleration." If you want to, you can turn to page 227 and review the sample chapter. Next, read the essay question and the answer. Then answer Questions 31-33 by writing Yes or No. Answer Question 34 if you want to practice writing essay answers.

Question:
Contrast how you figure speed and how you figure distance. In your answer make sure you include the two formulas.

Answer:
The formula $s = d/t$ is used to calculate speed. s stands for speed; d stands for distance; and t stands for time period. The amount of speed then is the distance traveled in some time period. The speed of a moving body is the rate at which it covers distance.

Distance, on the other hand, is how far something travels in a specific time period at a constant speed. The formula used to figure distance is: $d = s \times t$. The same factors are involved as in the formula for speed, but obviously their arrangement is far different.

31. Has the essay question been turned into a statement to begin the answer?
     If no, write a sentence that would be a good way to begin the answer.

32. Is the answer the kind required by the direction word in the question?

33. Does the answer include all the points required by the question?

34. (Optional) To practice taking essay tests, write your own answer to this question.
     Use your own paper. Then ask your instructor to judge your answer.

EXERCISE 13 ████████████████████████████████

The essay question below is based on the sample chapter "Speed, Velocity, and Acceleration." If you want to, you can turn to page 227 and review the sample chapter. Next, read the essay question and the answer. Then answer Questions 35-37 by writing Yes or No. Answer Question 38 if you want to practice writing essay answers.

Question:
Discuss the importance of knowing velocity of moving bodies.

Answer:
Velocity involves both speed and direction in which something is moving. Velocity can be defined as a specification of both its speed and the direction in which it's moving. If velocity is constant, acceleration is zero. Velocities are different if speed is the same but directions are different.

35. Has the essay question been turned into a statement to begin the answer?
    If no, write a sentence that would be a good way to begin the answer.

36. Is the answer the kind required by the direction word in the question?

37. Does the answer include all the points required by the question?

38. (Optional) To practice taking essay tests, write your own answer to this question.
    Use your own paper. Then ask your instructor to judge your answer.

SUMMARY ███████████████████

1. To prepare for a test, follow a sequence of steps:
    a. Find out as much about the test as you can.
    b. Start preparing at least one week before the test.
    c. Gather together all the material you need to study and review it, noting important information.
    d. Summarize the material by making an outline that combines all the information you need to learn.
    e. Recall and recite that material daily.

2. When taking a test, remember these general guidelines:
    a. Read all the test directions carefully and survey the entire test before answering any questions.
    b. Budget your time, so you'll be able to complete the whole test.
    c. Read the questions carefully and answer those you are sure of first. If there's no penalty for wrong answers, guess.
    d. Try to save time to review your answers before submitting your test.

OBJECTIVE QUESTIONS ███████████████

3. Objective tests are generally used to measure your knowledge of specific facts and details. An objective question has one correct answer. The most common kinds of objective test questions are true-false, multiple-choice, matching, and short-answer.

4. To answer a true-false question, you must decide if a statement is true or false.
    a. When a true-false question contains an absolute, the statement is usually false. Absolutes are words like *always, never, all*, and *none* — words that say there are no exceptions. When a true-false statement contains a qualifier, like *generally* or *often*, the statement is usually true.
    b. To simplify a true-false question that contains a double negative, try crossing out both negatives. Then decide if the statement is true or false.
    c. If any part of a true-false question is false, the whole statement is false.

5. In a multiple-choice question you are given a number of possible answers. You must choose the correct answer from the choices given.

    a. If you're not sure of the answer to a multiple-choice question, use a process of elimination. First, read all the answer choices. Then eliminate the choices that you know are wrong.

    b. You can usually eliminate choices that contain absolutes, choices that are too general, and choices that give only part of the answer.

    c. The longest answer choice in a multiple-choice question is often the most complete. If so, it is probably the correct answer.

    d. To know that "all of the above" is the right answer in a multiple-choice question, you need only be sure that two other choices are correct.

6. To answer a matching question, you must match each item in one list with an item from another list. Read the directions carefully to see if an item can be used more than once.

7. On a short-answer test, you supply the answers to the questions. Sometimes the number of lines is a clue to the number of words in the answer. The word *an* before a blank line tells you that the answer begins with a vowel sound.

ESSAY QUESTIONS

8. Essay tests usually call for detailed, written answers. There's no one correct answer to an essay question. An instructor reads the answer and judges whether it is complete, accurate, and well-organized.

9. It is essential to budget your time when you take an essay test. Read the directions and all the questions first.

10. Use a system of check marks to decide which questions to answer first. Put one check mark next to the questions with the highest point value, and two check marks next to the questions that you know you can answer quickly. The questions to answer first are the ones with three check marks; these are questions with a high point value and that you know you can answer quickly. The questions to answer second are the ones with two check marks; these are questions that you know you can answer quickly, but they don't have a high point value. Answer the questions with one check mark next, and the questions with no check marks last.

11. Set a time limit for each question, allowing more time for questions that have a higher point value. Keep an eye on the time as you are working, to make sure you'll have time to answer all the questions.

12. Pay particular attention to the direction words in an essay question. The direction word tells you what kind of answer is expected.

13. To answer an essay question, follow these steps:
    a. Plan your answer. Make a list of the important points, or write an outline for the answer. If you don't have time to answer the question, write the outline instead.
    b. Turn the question into a statement at the start of your answer.
    c. Write neatly and follow your outline.
    d. Leave space at the end of each answer so you'll have room if you decide to add more information later.
    e. Save some time to review your answers.

Additional Suggestions for Science Students

In introductory science courses, written tests are usually given. As stated earlier, they contain many objective questions, and they may contain some short essay questions. In addition, problems requiring mathematical computation and application of formulas may be included too, particularly on physics and chemistry tests.

In addition to performance on written tests, students can expect their lab performance to be evaluated. The manner in which a student approaches lab work and follows through in completing the experiments is observed. The quality of student's written work, the observations and conclusions noted for these experiments, is then definitely evaluated. In addition, lab tests are given periodically. For example, several stations may be set up in the lab, each with a different task or experiment to be completed by each student. Performance at each of the stations is evaluated. Final lab grades are often based on both the written work submitted for experiments completed throughout the term and on the practical lab tests. It's important to be most familiar with the scientific principles involved in order to perform well in the lab.

9

Using the Library

As a student, you'll often use the study areas in the library to prepare homework assignments and to review for tests. For some assignments, you'll also need to use the materials that are kept in the library. Knowing how to use the library and how to find books and periodicals is an important study skill.

APPLYING THIS SKILL

To practice applying this skill, you need:

1. This book
2. A pen or pencil

Take these items to a work area. Read the review section to make sure you understand the information you need in order to apply this skill. Then complete the exercises for Skill 9.

PERFORMANCE OBJECTIVE

When you have completed Skill 9, you will be able to use the card catalog and periodical indexes to find the reading material you need.

Using The Library

In a college library a great deal of space is used for tables, chairs, and study carrells. As a student, you'll often use these study areas to prepare homework assignments and to review for tests. In order to complete some assignments, you'll need to use the materials housed in the library too; therefore, you need to be familiar with the library's major areas.

Upon entering the library, you can easily detect the loan counter, sometimes called the circulation desk/counter. At that counter, library materials are checked out and returned. Sometimes reserve items are loaned there too, and library cards may be issued there. Books in the general collection are usually loaned the longest period of time, perhaps a month or so while other library material may be borrowed for only a week or two. Reserve items usually may not be checked out for more than a specified number of hours. Reference materials, newspapers, microforms, and the most recent issues of magazines/periodicals are for library use only.

The largest and perhaps most obvious section of the library housing material is the general collection of books on open shelves. Non-fiction or information books are arranged on the shelves according to the call numbers placed on the back of the books. Fiction books are arranged alphabetically by the authors' last names in a special section of open shelving. Some books are not on the open shelves but are on shelves or "stacks" in an area closed to the public. Books may be obtained from the area by filling out a "call slip", a form used to write the call number, author, title, publisher, and date of publication, and submitting it to the person designated. Back issues of magazines/periodicals are kept in an area closed to the public too. If the library has the magazine you want, you fill out a magazine request form for each item you need and submit it to the person in charge. Usually, paper issues of the magazines may be checked out of the library, or you can xerox articles found in them.

There is also a reference section in a library where reference books are shelved by call number. These books must be available at all times, so they may not be checked out. Typical reference books of which all science students should be familiar are the following: dictionaries of scientific terms; biographical dictionaries containing concise histories of the lives of noted scientists; yearbooks which contain up-to-date technical and scientific facts of the very latest year; encyclopedias usually bound in several volumes covering a wide range of information with topics arranged alphabetically; and reference manuals providing scientific and technical information sources. Usually, a reference librarian is available for assistance in the reference section.

Then in most libraries there is an area where the most recent issues of magazines/periodicals are arranged on open shelving for immediate perusal. Current newspapers are made available too. These magazines and newspapers may not be checked out of the library.

In addition to books, magazines, and newspapers, libraries contain instructional materials in several formats. Microfiche and microfilm contain printed material from various periodicals and publications. Usually these microform materials are kept in special steel cabinets. One may view them on readers arranged in an area of the library. In that area, special reader-printers, capable of producing photocopies of microform matter, are often available too.

Finally, in order to locate specific materials in the library, one must be familiar with two areas in which the key tools to libraries are housed. The major tools which you'll need to use are the card catalog and periodical indexes. The card catalog is a file in which entries on separate cards are arranged in alphabetical order, listing materials in the library by author, subject, and title. These are the three ways in which a book is usually listed in the card catalog. The same information may be available on a *computer*, but the same skills are needed for locating material when using a computer as when using the card catalog; therefore, the discussion here refers to the card catalog containing actual cards.

Sample author, subject, and title cards for a book of interest to science students are included at the end of this review. You may refer to them as you review the important points listed below for reading cards accurately.

Let's review now what you need to know in order to locate and read cards filed in the card catalog.

1. **Author Card** - a card which has the name of an author on the top line. Author cards form a list of all of the books by a single author. Author cards are filed alphabetically by the author's last name. If a book has more than one author, the catalog contains a separate card filed under each author's last name. Information on these cards includes the following:
   a. Call Number - letters, figures, symbols assigned to library materials to indicate their location; the year that a specific edition of the book was published is sometimes printed as the bottom line of the call number.
   b. Author's or authors' names
   c. Title of the book
   d. Place of publication
   e. Publisher
   f. Date of publication
   g. Paging
   h. Special features such as illustrations and bibliography/bibliographical notes

i. Tracings - entries listed on the bottom of the card which show the headings for cards entered in the catalog to help locate a given book. This means that for each heading listed as a tracing, there is a corresponding card filed in the catalog. There are three types of entries: author, subject, and title. The **subject headings** listed in the tracings are always identifiable by the Arabic numeral designation (l,2,3). Other entries such as joint authors, editors, compilers, and others connected with the book's publication as well as title and variations of the title are designated by Roman numerals (I,II).

Below is an author card for a book in a science area:

Author Card

```
QC
23          Beiser, Arthur
B4143           Modern technical physics / Arthur
1979        Beiser.— 3rd ed. — Menlo Park, Calif:
            Benjamin/Cummings Pub. Co.,
            c1979.
                866 p. :  ill. ; 24 cm.

                Includes index.
                ISBN  0-8053-0680-3


                1.  Physics.  I.  Title.


QC23.B4143   1979                            530
                                          78-31596
```

2. **Title Card** - a card which contains the same information as the author card with the exception that the title of the book is typed above the author's name. Title cards are filed alphabetically by the first major word in the title. If the title begins with *A, An,* or *The,* the card is alphabetized by the second word in the title.

3. **Subject Card** - a card which contains the same information as the author card with the exception that the subject heading is typed above the author's name. The heading is typed completely in **capital letters**. If a book treats more than one subject, additional subject cards may be used for the book. By looking under a subject heading in the card catalog, the number of books which the library has on that subject can be determined.

Below are title and subject cards listing the same book in the science area:

Title Card

```
                    Modern technical physics

Qc
23
B4143          Beiser, Arthur.
1979               Modern technical physics / Arthur
               Beiser. — 3rd ed. — Menlo Park, Calif:
               Benjamin/Cummings Pub. Co.,.
               c1979.
                   866 p. :  ill. ; 24 cm.

                   Includes index.
                   ISBN 0-8053-0680-3


                   1. Physics.  I.  Title

QC23.B4143  1979                                 530
                                            78-31596
                    B 011 855                LC MARC
```

Subject Card

```
                    PHYSICS.

QC             Beiser, Arthur.
23                 Modern technical physics / Arthur
B4143          Beiser. — 4th ed. — Menlo Park,
1983           Calif.: Benjamin/Cummings Pub. Co.,
               c1983
                   xv, 842 p.  :  ill. ; 25 cm..

                   Includes index.
                   ISBN 0-8053-0682-X


                   1.  Physics.  I.  Title.

SJCC DI          840413      840409
C000103            /SJG      A*      84-B679
                             84-12992
```

In addition to author, title, and subject cards, the card catalog also contains cross reference cards. They tell you to look under another subject to find cards for the books you need. On these cards you are directed to "see" or "see also." A **see** reference, for example, may direct you from a commonly-used term like BIO-LOGY-ETHICS to the subject heading used in the card catalog, BIOETHICS. **See also** references direct you from one subject heading like CHEMISTRY-HISTORY to a closely related subject heading(s) such as CHEMISTS. By looking under an additional subject heading(s), you may locate more material about a given topic.

When you need a book, fill out a call slip. Copy the following information from the card in the card catalog:

   a.   the call number
   b.   the title of the book
   c.   the author of the book
   d.   the publisher of the book
   e.   the year the book was published

Science students need to be aware when using the card catalog, that in addition to a variety of cards for resource books, cards for special dictionaries, encyclopedias, yearbooks, and reference manuals may be filed. Often, however, these materials are shelved in the reference section of the library and cannot be checked out for personal use. Also included in the card catalog may be cards for non-book materials such as phonograph records, film loops, tapes, and microfiche. These materials may be identified by the description on the cards and by colored banding or a stripe at the top of each card, coded according to the type of material listed. Science students are frequently directed to use both reference material and non-book material in addition to the books found in the stacks.

The second major tool you'll need to use to locate material in the library is the periodical index. A periodical index is a listing which locates periodical articles primarily by subject. One of the most frequently used indexes and the most basic index to general interest, non-technical periodicals is the **Readers' Guide to Periodical Literature,** usually called the **Readers' Guide.** This index is very wide in scope; it usually covers any subject one would expect to find in over one hundred and fifty popular, well-known magazines which it indexes. When you know how to use the **Readers' Guide,** you can use almost any periodical index available.

A sample entry from the **Readers' Guide** is included at the end of this review. You may refer to it as you review the important points listed below for accurately locating and reading index entries. Specific reference is made to the **Readers' Guide** because knowing how to use this index enables you to use almost any index.

Let's review now what you need to know in order to locate and read index entries:

1. Paperback issues of the **Readers' Guide** are published every two weeks and are realphabetized or cumulated and republished into quarterly issues and into one large bound volume published at the end of each year. The year of each annual volume is printed on the back binding of the volume. At the front of each paperback issue and bound volume of the **Readers' Guide** and most other periodical indexes are the following:
    a. A list of abbreviations used for the titles of the periodicals included;
    b. A list of other abbreviations and symbols used in the entries;
    c. A sample entry with an explanation of its meaning.
    These aids should be consulted whenever a periodical index is used.

2. Articles in periodical indexes are indexed primarily by subject. The subject is designated by a main subject heading printed to the far left of the column in which articles about the subject are listed. Only the first word of the subject heading appears in capital letters. In addition, subheadings for particular aspects of a subject are sometimes provided. Subheadings are printed in the center of the column under the main subject heading. Only the first letter in subheadings is capitalized.

3. Entries are made using a standard format:
    a. Each entry under the subject heading begins with the title of the article. Only the first letter of the title is capitalized.
    b. The title is followed by the author's name (if known).
    c. If any special features are included in the article, such as bibliography (bibl), illustrations (il), or portraits (por), abbreviations for them are indicated after the author's name.
    d. Then the title of the periodical in which the article appears is designated. Usually, the title is abbreviated.
    e. Next, the volume number of the periodical is designated. Publishers usually assign a volume number to include several issues of a periodical.
    f. The volume number is followed by a colon which is immediately followed by the specific page(s) on which the article appears. When a "+" appears after the last page number listed, the article is continued on some further page(s) in the periodical.
    g. Finally, the date of the issue in which the article appears is specified.

The following is a sample entry from the **Readers' Guide**. Arrows (→) indicate the subject heading, subheading, and the specific entry which is enlarged below:

Subject Heading →   **Gasoline**
    Subheading →   **Lead Content**
          Entry →   Senate considers lead gasoline ban. E. Marshall. *Science*
                225:34-5 Jl 6 '84
                   Octane rating
            Higher octane may not be worth the cost. D. H. Dunn.
              Bus Week p130 S 17 '84
          Gastineau, Mark
                    about
            NO! NO! NO! Gastineau. B. Newman. il pors *Sports Illus*
              61 Sp Issue:46-50 S 5 '84

Source: **Readers' Guide to Periodical Literature.** Volume 84, Number 13, October 25, 1984, pg. 62.

Entry enlarged:

          1   2   3 4   5   6
Senate considers lead gasoline ban. E. Marshall. *Science* 225: 34-5
              7
       Jl 6 '84

1. Title of article
2. Author
3. No special feature is indicated, but if it had been, its abbreviation would have been listed after the author's name and before the name of the periodical.
4. Title of the periodical in which the article appears
5. Volume number of the periodical
6. Pages on which the article appears
7. Date of the issue in which the article appears

When you need to find an article in a periodical, copy the following information from the periodical index:

   a. the title of the article
   b. the title of the periodical
   c. the volume number, if any
   d. the page numbers on which the article appears
   e. the date of the issue

Science students should know that in addition to the **Readers' Guide to Periodical Literature,** other specialized indexes are available for use. Often, periodicals more technical and scholarly in nature than those in the **Readers' Guide** are indexed in specialized indexes. The periodicals indexed contain articles specifically addressing science topics. Generally, specialized indexes are organized in the same way as the **Readers' Guide.** In the front of the index is a page listing the periodicals indexed and their abbreviations used within entries. There is also a page in the front of the index listing abbreviations for the terms frequently used within entries, as well as a sample entry. Since the entries in these specialized indexes follow the same basic format as that used for entries in the **Readers' Guide,** they are read basically in the same manner.

An example of a specialized index listing articles of importance to science students is the **Applied Science and Technology Index.** This index covers articles on aeronautics, automation, chemistry, construction, electricity, electronics, engineering, geology, machinery, physics, transportation, and related subjects. Below is a sample entry from this index:

Subject Heading ➞ **Radicals (Chemistry)**

Absolute rate constants for radical rearrangements in liquids obtained by muon spin rotation. P. Burkhard and others. bibl diags *J Phys Chem* 88:773-7 F 16 '84

An ESR investigation of the $C^{17}O$ cation radical isolated in a neon matrix at 4 K. L. B. Knight, Jr. and J. Steadman. bibl *J Am Chem Soc* 106:900-2 F 22 '84

Free redicals in lipod bilayers: new probes of lipid radical dynamics. N.A. Porter and others. bibl diags *J Am Chem Soc* 106:813-14 F 8 '84

Entry ➞ Gaseous nitrate radical: possible nighttime atmospheric sink for biogenic organic compounds. A. M. Winer and others. bibl diags *Science* 224:156-9 Ap 13 '84

Geometry of the CH^2OR radical in X-irradiated crystals of methyl B-D-galactopyranoside: an ESR/ENDOR study. W. A. Bernhard and others. bibl diags *J Phys Chem* 88:1317-20 Mr 29 '84

Highly specific reciprocal methyl/hydrogen transfer reactions preceeding some unimolecular dissociations of crowded enol cation radicals in the gas phase as examples for conformationally controlled processes. S. E. Biali and others. bibl diags *J Am Chem Soc* 106:496-501 F 8 '84

From: **Applied Science and Technology Index,** Volume 72, Number 8, September, 1984, pg. 394.

The subject heading is Radicals (Chemistry). The article indicated is about that subject heading. Notice there is no subheading provided in this example. The title of the article is "Gaseous Nitrate Radical: Possible Nighttime Atmospheric Sink for Biogenic Organic Compounds." The author's name is A. M. Winer and others. There are two special features: a bibliography and diagrams. The name of the periodical in which this article appears is **Science.** The volume number of the periodical is 224; the pages on which the article appears are 156-159; and the date of the issue in which this article appears is April 13, 1984.

An additional specialized index listing articles in the science areas is the **Education Index** which covers all phases of education. Another specialized index listing articles which focus on science topics is the **General Science Index**. This index covers articles on astronomy, biology, botany, chemistry, earth science, food and nutrition, genetics, mathematics, medicine, oceanography, physics, physiology, pyschology, zoology, and related subjects.

In all periodical indexes there may be **see** and **see also** references following some subject headings and subheadings. Below these words, one or more additional subject headings are listed. **See** references direct you to the subject heading which is used instead of the one under which you looked. **See also** references direct you to related subject headings under which additional material may be found. **See** and **see also** references are used in both the card catalog and periodical indexes in the same manner as they are used in textbook indexes which you reviewed earlier.

EXERCISES

There are 10 exercises for Skill 9. In Exercises 1-5, you will answer questions about cards from the card catalog. In Exercises 6-10, you will answer questions about entries from various periodical indexes. A pointed finger (☞) marks the entry about which these questions are asked. Write your answers in the spaces provided in your book. You can check your answers with the answer key at the back of this book.

EXERCISE 1 ████████████████████████████████████

A card from the card catalog is shown below. Use it to answer Questions 1-6.

1.   Fill out the call slip for this book.

2.   What kind of card is this?   CALL SLIP
      (author, title, or subject)

3   Where was the book published?

4.   How many pages are in the book?

5.   What special features are included
      in this book? (If none are indicated
      on the card, write "none.")

6.   What is/are the subjects under
      which this book is listed?

```
OC          Marion, Jerry B.
21.1            Principles of physics / Jerry B.
M366        Marion and William F. Hornyak. —
            Philadelphia :  Saunders College Pub.,
            c1984.
                xvii, 772 p,  10 : ill.   ; 27 cm. —
            (Saunders golden sunburst series)

                Includes index.
                ISBN 0-03-049481-8

                1.  Physics.  I.  Hornyak, William F.
            (William Frank), 1922-  II. Title.
```

EXERCISE 2 ■■■■■■■■■■■■■■■■■■■■■■■■■

A card from the card catalog is shown below. Use it to answer Questions 7-12.

7.   Fill out the call slip for this book.

8.   What kind of card is this?   CALL SLIP
      (author, title, or subject)

9.   Where was the book published?

10.   How many pages are in the book?

11.   What special features are included
      in this book? (If none are indicated
      on the card, write "none.")

12.   What is/are the subjects under
      which this book is listed?

```
              CHEMISTRY

OD        McQuarrie, Donald A.
31.2          General chemistry / Donald A.
M388      McQuarrie, Peter A.  Rock. — New York :
          W. H. Freeman, c1984.
              xviii, 1063 p, (80),  :  ill.  27 cm.

              Includes index.
              ISBN 0-7167-1499-x


              1.  Chemistry.   I. Rock, Peter A.,
          1939-  II.  Title.
```

EXERCISE 3 ▰▰▰▰▰▰▰▰▰▰

A card from the card catalog is shown below. Use it to answer Questions 13-18.

13. Fill out the call slip for this book.

14. What kind of card is this?
    (author, title, or subject)

15. Where was the book published?

16. How many pages are in the book?

17. What special features are included
    in this book? (If none are indicated
    on the card, write "none.")

18. What is/are the subjects under
    which this book is listed?

CALL SLIP

```
          BIOLOGY

QH        Rahn, Joan Elma, 1919-
308.2         Biology; the science of life / Joan E.
P34       Rahn. - 2d ed. - NewYork : Macmillan,
          c 1980

              xi, 673 p., (6) leaves of plates :
              ill. ; 28 cm.

              Includes bibliographies and index
              ISBN 0-02-397620-9

              1. Biology.  I.  Title.
```

EXERCISE 4 ████████████████████████████

A card from the card catalog is shown below. Use it to answer Questions 19-24.

19.   Fill out the call slip for this book.

20.   What kind of card is this?   CALL SLIP
       (author, title, or subject)

21.   Where was the book published?

22.   How many pages are in the book?

23.   What special features are included
       in this book? (If none are indicated
       on the card, write "none.")

24.   What is/are the subjects under
       which this book is listed?

```
                    Physics, for scientists and
                         engineers.

        OC         Serway, Raymond A.
        23             Physics, for scientists and
        S458       engineers / Raymond A. Serway. —
                   Philadelphia : Saunders College Pub.,
                   c1982.
                       xviii, 883 p,  vi,   :  ill.  ; 26
                   cm. — (Saunders golden sunburst
                   series)

                       Includes bibliographical references
                   and index.
                       ISBN 0-03-057903-1

                       1.  Physics.  I.  Title
```

EXERCISE 5 ■■■■■■■■■■■■■■■■■■■■■■■■■

A card from the card catalog is shown below. Use it to answer Questions 25-30.

25. Fill out the call slip for this book.

26. What kind of card is this?
     (author, title, or subject)

27. Where was the book published?

28. How many pages are in the book?

29. What special features are included
     in this book? (If none are indicated
     on the card, write "none.")

30. What is/are the subjects under
     which this book is listed?

CALL SLIP

```
                CHEMISTRY

QD        Ebbing, Darrell D.
31.2          General chemistry / Darrell D.
E22       Ebbing ; consulting editor, Mark S.
          Wrighton. — Boston : Houghton Mifflin
          Co., c1984.
              xxiv, 970p, (98) ,   (8) p. of plates
          : ill. (some col.)  : 26 cm.

              Includes index.
              ISBN 0-395-31489-5

              1.  Chemistry.   I.  Wrighton, Mark S.
          1949-   II. Title.
```

EXERCISE 6 ████████████████████████████

Use the section from the index on this page to answer Questions 31-40.

31.   Under what main subject heading is the article listed?

32.   Under what subheading of the subject is the article listed?
       (If no subheading is provided, write "none.")

33.   What is the title of the article?

34.   Who wrote the article?

35.   What special features (if any) are included in the article?
       (If none are indicated, write "none.")

36.   In what periodical did the article appear? Write the
       abbreviation for the title if you don't know the
       periodical represented.

37.   In what volume of the periodical
       did the article appear?

38.   On which page(s) in the periodical
       did the article appear?

39.   What form of punctuation separates
       the volume number from the page
       numbers in the entry?

40.   What is the complete date (do not
       abbreviate) of the issue in which
       the article appeared?

Plant growth
            See also
        Tree rings
Plants
            See also
        Aquatic plants
        Geographical distribution of animals and plants
            Disease and pest resistance
☞   Cultured cells of white pine show genetic resistance to
        axenic blister rust hyphae. A. M. Diner and others.
        bibl il *Science* 224:407-8 Ap 27 '84

Source: **Applied Science and Technology Index**. September 1984,
pg. 367.

EXERCISE 7 ████████████████████████████████

Use the section from the index on this page to answer Questions 41-50.

41.   Under what main subject heading is the article listed?

42.   Under what subheading of the subject is the article listed?
      (If no subheading is provided, write "none.")

43.   What is the title of the article?

44.   Who wrote the article?

45.   What special features (if any) are included in the article?
      (If none are indicated, write "none.")

46.   In what periodical did the article appear? Write the
      abbreviation for the title if you don't know the
      periodical represented.

47.   In what volume of the periodical
      did the article appear?

48.   On which page(s) in the periodical
      did the article appear?

49.   What form of punctuation separates
      the volume number from the page
      numbers in the entry?

50.   What is the complete date (do not
      abbreviate) of the issue in which
      the article appeared?

Chemical symbols *See* Chemistry--Notation
Chemicals
    Chemical of the month. C. H. Beach. See occasional
        issues of Journal of Chemical Education
                Labeling, marking, etc.
    What's on the label? C. Quailey. il *Sch Arts* 83:35 Ja
        '84
                Safety measures
☞   Chemistry workshops to prolong the lives of your favorite
        janitors. J. N. Aronson. *J Chem Educ* 60:1036-7
        D '83

Source: **Education Index**. July 1983-June 1984, pg. 177.

EXERCISE 8 ████████████████████████████

Use the section from the index on this page to answer Questions 51-60.

51. Under what main subject heading is the article listed?

52. Under what subheading of the subject is the article listed?
    (If no subheading is provided, write "none.")

53. What is the title of the article?

54. Who wrote the article?

55. What special features (if any) are included in the article?
    (If none are indicated, write "none.")

56. In what periodical did the article appear? Write
    the abbreviation for the title if you don't know
    the periodical represented.

57. In what volume of the periodical
    did the article appear?

58. On which page(s) in the periodical
    did the article appear?

59. What form of punctuation separates
    the volume number from the page
    numbers in the entry?

60. What is the complete date (do not
    abbreviate) of the issue in which
    the article appeared?

Hydrogen
    Isotopes
        See also
    Tritium
    Spectra and spectroscopy
☞   Observations of Lyman -a emissions of hydrogen and
        deuterium. J. L. Bertaux and others. bibl f il *Science*
        225:174-6 Jl 13 '84

Source: **Readers' Guide to Periodical Literature**. Volume 84,
Number 13, October 25, 1984, pg. 73.

EXERCISE 9 ██████████████████████████████████████

Use the section from the index on this page to answer Questions 61-70.

61.   Under what main subject heading is the article listed?

62.   Under what subheading of the subject is the article listed?
      (If no subheading is provided, write "none.")

63.   What is the title of the article?

64.   Who wrote the article?

65.   What special features (if any) are included in the article?
      (If none are indicated, write "none.")

66.   In what periodical did the article appear? Write the
      abbreviation for the title if you don't know the
      periodical represented.

67.   In what volume of the periodical
      did the article appear?

68.   On which page(s) in the periodical
      did the article appear?

69.   What form of punctuation separates the
      volume number from the page
      numbers in this entry?

70.   What is the complete date (do not
      abbreviate) of the issue in which
      the article appeared?

Biology
>   *See also*
>   Clones (Biology)
>   Cryobiology
>   Developmental biology
>   Embryology
>   Evolution
>   Information storage and retrieval systems —
>   Biological use
>   Natural history
>   **Classification**
> ☞   Taxonomy: what's in a name? C. J. Cole. il *Nat Hist*
>   93:30+ S '84

Source: **Readers' Guide to Periodical Literature**. Volume 84,
Number 13, October 25, 1984, pg. 20.

EXERCISE 10

Use the section from the index on this page to answer questions 71-80.

71. Under what main subject heading is the article listed?

72. Under what subheading of the subject is the article listed?
     (If no subheading is provided, write "none.")

73. What is the title of the article?

74. Who wrote the article?

75. What special features (if any) are included in the article?
     (If none are indicated, write "none.")

76. In what periodical did the article appear?
     Write the abbreviation for the title if you
     don't know the periodical represented.

77. In what volume of the periodical
     did the article appear?

78. On which page(s) in the periodical
     did the article appear?

79. What form of punctuation separates
     the volume number from the page
     numbers in the entry?

80. What is the complete date (do not
     abbreviate) of the issue in which
     the article appeared?

Heredity of diseases
☞   Gene therapy method shows promise. G. Kolata. *Science*
      223:1376+ Mr 30 '84

Source: **Applied Science and Technology Index.**
September, 1984, pg. 230.

SUMMARY

1. To use the library efficiently, you need to be familiar with its major sections. These are the major sections found in most libraries:

   a. The **circulation desk**, or **main desk**, is where books are checked out and returned.

   b. The **stacks** contain the collection of circulating books that can be taken out of the library.

   c. The **reserve section** holds the library materials in which instructors have assigned reading for students.

   d. The **reference section** contains reference books, such as encyclopedias and almanacs.

   e. The **periodical section** is where the current issues of magazines, journals, and sometimes newspapers are kept.

   f. The **microform section** stores photographic reproductions of printed materials, such as back issues of periodicals.

2. Every book in the library is assigned a **call number**. The call number appears on the spine of the book. Books are arranged on the library shelves according to their call numbers. To find the call number of a book, use the card catalog.

3. The **card catalog** is an alphabetic index of all the books in the library. It contains at least three cards for each nonfiction book in the library: an author card, a title card, and at least one subject card. If you know the title of the book you need, it's probably fastest to look for the title card. If you know the author's name, but not the title, look for an author card. To find out what books the library has on the subject you are interested in, look at the subject cards.

4. Some libraries file author, title, and subject cards separately. In other libraries, author cards and title cards are filed together, and a separate file is used for subject cards. The card catalog is arranged in this way:

   a. **Author cards** are filed alphabetically, by the author's last name. If a book has more than one author, the catalog contains a separate card under each author's last name.

   b. **Title cards** are filed alphabetically by the first major word in the title. If the title begins with *A, An*, or *The*, the card is alphabetized by the second word in the title.

   c. **Subject cards** are filed alphabetically under the general subject of the book. Many books have more than one subject card.

d. A **cross-reference card** tells you to look under another subject to find cards for the book you need.

5. When you need a book, fill out a call slip. Copy the following information from the card in the card catalog:
    a. the call number
    b. the title of the book
    c. the author of the book
    d. the publisher of the book
    e. the year the book was published

6. Other useful information on a card in the card catalog includes the place of publication, the number of pages, and the special features in the book. Entries at the bottom of the card, called tracings, show all the subjects under which the book is listed in the card catalog.

7. A **periodical** is a work that is published on a regular basis, such as a daily newspaper, a weekly magazine, or a monthly journal.

8. A **periodical index** is a reference book that lists articles from periodicals. There are many different periodical indexes; many deal with specific subjects. One of the most widely used indexes is the **Readers' Guide to Periodical Literature**. **The Readers' Guide** is an index of articles from a large number of general interest periodicals.

9. Entries in the **Readers' Guide** are listed alphabetically. You can find the entry for an article by looking under the subject heading and, often, a subheading. Or you can look under the author's last name.

10. When you need to find an article in a periodical, copy the following information from the periodical index:
    a. the title of the article
    b. the title of the periodical
    c. the volume number, if any
    d. the page numbers on which the article appears
    e. the date of the issue.

11. An entry in a periodical index will also show the name of the author (if there is one) and any special features included in the article, such as illustrations or maps.

Additional Information for Science Students
Library Resources for Studying Science

In most libraries, the reference section has a number of reference books that can be very useful to science students. You will probably find several dictionaries of scientific terms, some general and others specific to one field of science. The reference section may also have a biographical dictionary, which gives brief accounts of the lives of noted scientists. Encyclopedias of scientific subjects usually include several volumes; a wide range of scientific, mathematical, and technical topics are arranged alphabetically. Yearbooks or almanacs contain up-to-date technical and scientific facts and discoveries.

Several of the most widely used reference books in science are listed below. To find out which reference books your library has, you should ask the reference librarian for help.

Dictionaries

Bynum, W. F., E. J. Browne, and R. Porter (ed.). **Dictionary of the History of Science**. Princeton, New Jersey: Princeton University Press, 1981.

Daintith, J. (ed.). **A Dictionary of Physical Sciences**. New York: PICA Press, 1976.

Hawleye, G. G. **The Condensed Chemical Dictionary**. Tenth ed. New York: Van Nostrand Reinhold, 1981.

Lapedes, D. N. (ed.). **McGraw-Hill Dictionary of Physics and Mathematics**. New York: McGraw-Hill, 1978.

Tootill, E. (ed.). **The Facts on File Dictionary of Biology**. New York: Facts on File, 1981.

Encyclopedias

Asimov, I. **Asimov's Biographical Encyclopedia of Science and Technology**. Garden City, New York: Doubleday, 1982.

Parker, S. P. (ed.). **McGraw-Hill Encyclopedia of Science and Technology**. Fifth ed. (15 volumes) New York: McGraw-Hill, 1982.

Yearbooks

Parker, S. P. (ed.). **McGraw-Hill Yearbook of Science and Technology**. New
York: McGraw-Hill, 1984.

Reference Manuals

Chen, C. C. **Scientific and Technical Information Sources**. Cambridge, Mass.:
The MIT Press, 1977.

There are many journals and other periodicals that deal with scientific subjects. Some examples
are: **BioScience, Chemical and Engineering News, Journal of Chemical Education, Journal
of Physics, Microchemical Journal, Science,** and **Scientific American**. Some of these periodicals, such as **Science,** are included in the **Readers' Guide to Periodical Literature**. Others are
indexed only in specialized periodical indexes.

Sample Chapters:
The Sciences

Nine sample chapters from science textbooks are printed on the pages that follow. The sample chapters are to be used in completing the exercises for Skills 2, 5, 6, and 8. Each chapter is based on actual textbook material.

Chapter 12
Algae

What are the four common traits that all algae share?

What is the basis for color and cell structure for blue-green algae?

How do blue-green algae reproduce and where do they live?

Among the algae are some of the oldest and most primitive types of plant life. They vary in size tremendously. Some are microscopic in size. Others, such as the large seaweeds, may equal the height of a tall tree. One major type of algae is the blue-green algae. In this chapter you will learn about the nature of algae in general. You will also learn specific characteristics of the blue-green algae.

What Are Algae Like?

Four general traits. All algae share four traits. First, algae possess chlorophyll and carry on photosynthesis. *Chlorophyll* can be defined as the green pigment needed for metabolic processes in plants. *Photosynthesis* is the process by which light energy is absorbed and then converted to the chemical bond energy of glucose. Second, algae can live in both fresh and salt water. Third, algae are, in general, simple organisms lacking both specialized tissues for conducting water and plant organs such as roots, stems, and leaves. Fourth, they have complex reproductive processes. They certainly must be considered successful forms of life.

Importance to aquatic communities. Many people do not realize the importance of algae in an aquatic environment. Algae, however, are the primary food producers for other forms of acquatic life, and they serve as an important source of oxygen for them. Algae consume carbon dioxide from the air or water. Using energy from the sun, they combine the carbon dioxide with a water molecule. The result is glucose sugar and oxygen. Algae then supply oxygen as well as food to other forms of life.

Blue-Green Algae

Basis for color. Blue-green algae contain blue pigment. In combination with chlorophyll, which is green in color, the blue pigment gives algae their blue-green color. Some kinds of blue-green algae also contain a red pigment. Red pigment, in combination with other pigments, creates a near black color in some "blue-green" algae.

Cell structure. Blue-green algae are organized in a simple manner. They are the simplest in organization of all algae in several ways. First, there isn't a definite nucleus in a single cell. As a result, DNA is scattered throughout the cell. *DNA* is the material that carries the genetic information for reproduction. Second, chlorophyll in blue-green algae is usually attached to membranes. There are no chloroplasts to contain the chlorophyll. Third, single cells are often arranged in chains or filaments as shown in Figure 12-1. Fourth, there seems to be little division of labor among the cells. Each cell is like every other cell of the group. Fifth, groups of cells are often enclosed within a protective jelly-like layer.

Figure 12-1. **Blue-Green Algae**

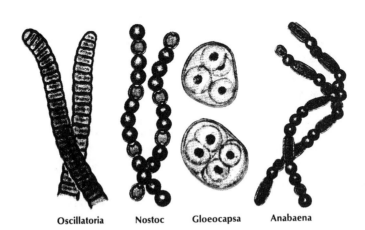

Oscillatoria   Nostoc   Gloeocapsa   Anabaena

From: Raymond F. Oram, Paul J. Hummer, Jr., and Robert C. Smoot, **Biology: Living Systems** (Columbus, Ohio: Charles Merrill Pub. Co., 1976), p. 314. Reprinted by permission of the publisher.

How blue-green algae reproduce. Blue-green algae reproduce asexually (without mating) through division of a single cell. New cells may be formed along a chain or filament through simple fission. *Fission* is reproduction in which a one-celled organism divides into two one-celled organisms. Occasionally, filaments are broken apart. Each section produces a new series of cells by fission. One kind of blue-green algae called Anabaena produces specialized cells that act as spores. These spores can develop into new filaments.

Habitat. Blue-green algae are commonly found in ponds, streams, and moist places on land. During hot summer months they multiply rapidly. As a result, public water supplies must be treated regularly during this season. Treatment is necessary to prevent rapid reproduction of blue-green algae and other microorganisms.

Summary

Algae, plants with common characteristics, are important to other forms of life in the aquatic environment. Blue-green algae contain blue pigment. The blue pigment combines with chlorophyll, green in color, to produce a blue-green color. Of all algae, blue-green algae are simplest in terms of organization and reproduction. They are commonly found in ponds, streams, and moist places on land.

Exercises

1. Describe the general traits of algae.
2. What is DNA?
3. How do blue-green algae reproduce?
4. Where would you find blue-green algae?
5. Describe the cell structure of blue-green algae.

Suggestion for Further Reading

Volpe, E. Peter. **Biology**, third edition. Dubuque, Iowa: Wm. C. Brown Co., 1983.

Chapter 13
Protozoa

After reading this chapter you will be able to:

1. describe protozoa in general
2. discuss the two types of protozoa--amoebae and ciliates

Protists are organisms that have characteristics similar to both plants and animals, but they are not animals or plants. They are mostly one-celled and belong to Kingdom Protista. Protozoa is one large group of protists. In this chapter you will learn about protozoa in general and more specifically about two common types of protozoa.

Protozoa in General

All protozoa share certain characteristics. First, they are one-celled, nongreen, animal-like protists. Second, they can move about and are classified according to how they move. Third, protozoa must secure their own food and are specially equipped to do so. Most protozoa actively trap their food. They get it from the surrounding water and consume it in chunks. Some protozoa, however, are parasites. They live directly off other organisms.

Amoebae

Forming pseudopodia. One of the simplest protozoans is the amoeba. Amoebae are unique in how they move. To move, they develop pseudopodia. *Pseudopodia* are false feet formed by flexible plasma membranes and cytoplasm. The formation of pseudopodia involves certain stages. Initially, the cytoplasm is in a fixed, gelatinous state. Then, chemical changes cause part of the cytoplasm to become a fluid. This fluid, which is like a true solution, is known as a sol. Cytoplasm in a sol condition can flow freely. Next, cytoplasm flows into the cell membrane, extending it to form a pseudopodium. Then this cytoplasm changes back to a gel. Finally, other areas of the cytoplasm undergo the same sol-gel transformation forming new pseudopodia. New pseudopodia are formed and others disappear. In this way, the shape of amoebae changes constantly. See Figure 13-1.

Figure 13-1. **Movement of the Amoeba**

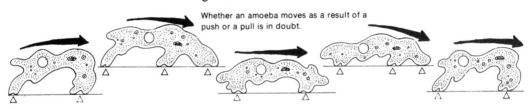

Whether an amoeba moves as a result of a push or a pull is in doubt.

From: Raymond F. Oram, Paul J. Hummer, Jr., and Robert C. Smoot, **Biology: Living Systems** (Columbus, Ohio: Charles Merrill Pub. Co., 1976), p. 327. Reprinted by permission of the publisher.

Food-getting structures. Pseudopodia are used in getting food as well as in moving about. Pseudopodia engulf particles of food. The food particles enter the cytoplasm where they are digested. You can see the food vacuoles in which digestion takes place. See Figure 13-2.

Figure 13-2. **Food-Getting Structures of the Amoeba**

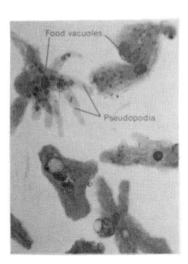

Habitat. The kind of amoeba studied in the laboratory is commonly found in ponds and other bodies of water. This type of amoeba is harmless. Some amoebae, however, are found in contaminated water. They can cause disease and are quite dangerous to humans.

Ciliates

General description. Ciliates, another common protozoa, have several characteristics. First, they are one-celled protozoa like amoebae. Second, they have a definite shape with a stiff covering bordering their cells, unlike amoebae. The stiff covering which borders cells of ciliates is called a *pellicle.* Third, on their surfaces they have hair-like cilia. *Cilia* are hair-like structures used for getting food and for moving about.

Paramecium. One group of ciliates often studied by biology students is the paramecium, which has certain characteristics. First, a paramecium is a complex organism in which different parts of the cytoplasm do different jobs. Second, a paramecium has two types of nuclei within a single cell. There is a large micronucleus which controls the basic activities of the cell. In addition, there's a smaller micronucleus which is involved in reproduction. Both nuclei help control large, complex ciliates. See Figure 13-3.

Figure 13-3. **Paramecium**

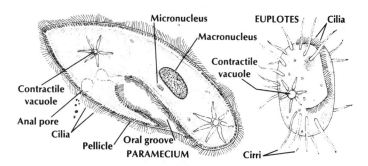

Courtesy of Roger K. Burnard, Coordinator, Bio-Learning Center, Ohio State University

Third, a paramecium moves by beating the cilia on its surface. The cell glides rapidly through water. Usually a paramecium rotates as it moves. See Figure 13-4.

Figure 13-4. **Rotating Movement**

From: Raymond F. Oram, Paul J. Hummer, Jr., and Robert C.
Smoot, **Biology: Living Systems** (Columbus, Ohio: Charles
Merrill Pub. Co., 1976), p. 328. Reprinted by permission of the
publisher.

Summary

Protozoa are one-celled, nongreen, animal-like protists. The amoeba is one of the simplest protozoans. Amoebae form pseudopodia or false feet used for moving about and for getting food. More complex protozoa are ciliates. Ciliates have hair-like cilia on their surfaces used for getting food and for moving about. Paramecium is a familiar ciliate. There is division of labor within its cell. Different parts of its cytoplasm do different jobs.

Exercises

1.   How does the amoeba move?
2.   Describe the sol-gel transformation in the amoeba.
3.   What are cilia? How are they used?
4.   Why do you think two nuclei may be necessary in a paramecium?

Suggestion for Further Reading

Guttman, Burton S. and Hopkins, Johns W. **Understanding Biology.**
     Orlando: Harcourt Brace Jovanovich, Inc., 1983.

Chapter 11
Origin of Genetics

What was Mendel's first experiment with genetics?
What is the law of dominance?
What was the result of his later experiments?

Genetics, a special branch of biology, is the science of heredity. To appreciate the significance of genetics, one must learn about Mendel (1822-1884) and his experiments. As a result of his work, the study of genetics was begun. In this chapter you will learn about Mendel's experiments and how they led him to the law of dominance.

Mendel's Initial Experiment

Purpose. Mendel experimented with pea plants to discover how traits are transmitted from parent to offspring. He wanted to arrive at a set of rules for the transmission of traits. He had already made certain observations about pea plants. First, he had seen that traits or visible characteristics are hereditary. *Hereditary* means that traits are passed on from generation to generation. Second, he had seen that many traits exist in one of two possible forms. For example, pea seeds are either round or wrinkled. The stems of pea plants are either tall or short.

Plants used in the experiment. For his experiment, Mendel used pure pea plants. *Pure plants* are those which after many generations of offspring have the same features as the parents. Pea plants reproduce sexually. There are both male and female sex organs in the same flower. Normally, male gametes (sex cells or pollen) fertilize eggs of the same flower. Pollen from the anther fertilizes an egg in an ovule. See Figure 11-1. The ovule becomes the seed. The ovary develops into a pod which protects the seeds. After many generations this process results in offspring with the same features as the parents or pure pea plants.

Figure 11-1. **Sexual Reproduction of Pea Plants**

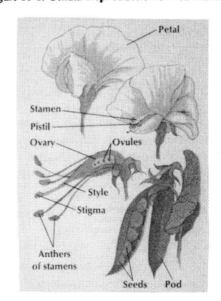

From: Raymond F. Oram, Paul J. Hummer, Jr., and Robert C. Smoot, **Biology: Living Systems** (Columbus, Ohio: Charles Merrill Pub. Co., 1976), p. 146. Reprinted by permission of the publisher.

Mendel's accomplishment. Mendel crossed pure plants having a particular trait with pure plants having the opposite trait. More specifically, he crossed pea plants which produced round seeds with pea plants which produced wrinkled seeds. To do this, he transferred the pollen of a plant with one trait to plants with the opposite trait, or opposite form of the trait. Pollen was obtained from round-seeded plants and transferred to wrinkle-seeded plants. The opposite procedure was also performed.

Results. Mendel found that in every case the parental cross (P) yielded offspring with round seeds only. The offspring of a parental cross are called the *first filial*, or F1, *generation*. There were no wrinkle-seeded plants in the F_1 generation.

Law of dominance. Mendel concluded that for each trait there is one form which "dominates" the other. A *dominant trait* is the trait which appears exclusively in the F_1 generation. For example, the trait for round seeds is the dominant trait. It "dominates" the trait for wrinkled seeds. The trait which disappears in the F_1 generation is called a *recessive trait*. In this case, the trait for wrinkled seeds is the recessive trait. Mendel generalized this result and formulated the law of dominance. The Law of Dominance states that one trait, the dominant trait, dominates or prevents the expression of the recessive trait.

Further Experimentation

F_2 generation. The lack of wrinkle-seeded plants in the F_1 generation impressed Mendel. Therefore, he experimented further. He then crossed members of the F_1 generation. Some F_1 round-seeded plants were crossed with other F_1 round-seeded plants. The offspring of the F_1 generation cross are called the *second filial,* or F_2, *generation.*

Results. Mendel found a ratio of 3:1; there were close to three round-seeded plants to each wrinkle-seeded plant. Of the 7,324 offspring produced, 5,474 plants had round seeds and 1,850 had wrinkled seeds. In this case, and in additional experiments with other pairs of traits, Mendel regularly obtained a 3:1 ratio in the F_2 generation.

Summary

Mendel's experiments provided the basis for the study of genetics. Mendel had observed that with pea plants visible characteristics or traits are hereditary. In addition, hereditary traits existed in one of two possible forms. He then set out to study the transmission of traits from parent to offspring. His first experiment involved crossing pure plants having one trait with pure plants having the opposite trait or a different form of a certain trait. All of the offspring of this cross (F_1 generation) had only one form of the trait. Mendel concluded that for each trait there is one form which "dominates" the other (opposite) form. This finding led to the law of dominance. He found there was a 3:1 ratio in the F_2 generation. The "dominant" trait appeared three times for each time the opposite trait appeared. The trait missing in the F_1 generation, the recessive trait, appeared in one fourth of the F_2 plants.

Exercises

1. How were the results of Mendel's parental crosses unusual?
2. How did Mendel interpret the results?
3. What is the "law of dominance"?
4. What happened when Mendel crossed members of the F_1 generation?

Chapter 2
Speed, Velocity, and Acceleration

After reading this chapter you will be able to:

   describe the basic theories of speed, velocity, and acceleration
   write the mathematical formulas for computing them

Physics is the most basic of the natural sciences. It is the science that deals with fundamental aspects of energy and nonliving matter. The concept of energy involves physical principles and measuring units with which we need to be familiar.

The basic notions of speed, velocity, and acceleration are fundamental to understanding the concepts of forces and energy. This chapter is intended to introduce you to speed, velocity, and acceleration.

Speed

Speed and its formula. Speed is the most commonly used rate. The amount of *speed* is a distance traveled in some period of time. The following formula can be used to express this:

$$\text{speed} = \frac{\text{distance traveled}}{\text{time period}} \text{ or } s = \frac{d}{t}$$

Consider an example. A runner completes a 100-yard dash in 10 seconds. His speed would be:

$$s = \frac{100 \text{ yd}}{10 \text{ sec}} = \frac{10 \text{ yd}}{\text{sec}}$$

Each automobile has a speedometer. It continuously shows the speed of the auto. An American automobile's speedometer shows the speed in miles/hour.

Distance. Knowing what the speedometer reads allows you to figure the distance a car will travel in a certain time period. Consider an example. Suppose the speedometer reads 60 miles/hour, and this speed stays the same. In a time period of three hours, the car will travel a distance of 180 miles. The formula used to figure distance is:

$$\text{distance} = \text{speed} \times \text{time or } \mathbf{d = st}$$

In the example stated above:

$$\mathbf{d} = 60 \text{ mi} \times 3 \text{ hrs or } 180 \text{ miles}$$

Velocity

The concept of speed avoids direction. Many situations, however, require one to know the direction in which an object travels. It is just as important as knowing the distance. For example, a pilot may need to fly from New York to Chicago. He'll want to know both the speed and direction of the plane. The *velocity* of a body is a specification of both its speed and the direction in which it is moving. A car, for example, may travel due north at 50 miles/hour. Another car may travel due east at 50 miles/hour. They would travel at the same speed. Their velocities, however, would be different. When direction is implied, velocity is the term to use.

Acceleration

Acceleration response. Every driver knows that he must press down the accelerator pedal to increase the speed of his car. All cars do not, however, respond in the same way to a depression of an accelerator. Some will change speed much faster than others. The *acceleration response* refers to the time it takes to achieve a certain speed starting from rest. If you compare the acceleration of two cars, the one reaching a specified speed in the least time has the greater acceleration. Acceleration can be expressed by the formula:

$$\text{acceleration} = \frac{\text{change in speed}}{\text{time required for change to occur}}$$

Direction and acceleration. Direction can be part of acceleration. Acceleration then refers to a change in velocity rather than speed. You can say then that an object can be accelerated by changing either (or both) the speed and direction.

Consider an example. A car travels at a constant speed of 50 miles/hour. It is traveling on a circular track. The car is accelerating because its direction is constantly changing. See Figure 2-1.

Figure 2-1. **Direction and Acceleration**

A car is traveling at a constant speed of 50 miles/hour on a circular track. It is accelerating because its direction is constantly changing.

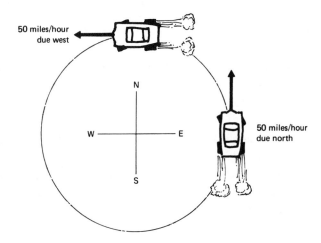

50 miles/hour
due west

50 miles/hour
due north

From: Joseph Priest, **Energy for a Technical Society**, ©1975, Addison-Wesley Pub. Co., Inc., Reading, Massachusettes, p. 30. Reprinted with permission.

Summary

The *speed* of a moving body is the rate at which distance is being covered by the body. The *velocity* of a body is a specification of both its speed and the direction in which it is moving. The *acceleration* of a body is the rate at which its speed or velocity changes with time. An object, therefore, can be accelerated by changing either (or both) the speed and direction.

Topical Review

1. Distinguish between speed and velocity.
2. What is acceleration?

Exercises

1. Can a rapidly moving body have the same acceleration as a slowly moving one?
2. How much distance does a car traveling at 35 miles/hour cover in 20 minutes?
3. A ship travels 400 miles in 1 day and 3 hours. What is its average speed in miles/hour?
4. If two cars are traveling the same speed, could one be accelerating?
5. What is the velocity of a horse traveling 25 miles per hour for 5 laps on a race track?

For Further Understanding

The day-to-day condition of the stock market as characterized by the Dow Jones averages, for example, is of great interest to many. How could you use the ideas of velocity and acceleration to characterize the stock market?

Suggestion for Further Reading

Beiser, Arthur. **Physics**, third edition. Menlo Park, CA: Benjamin Cummings Publishing Co., Inc. 1982.

Chapter 4
Newton's Three Laws of Motion

Force is a word used in various ways. It always, however, implies a stimulus for motion of some kind. Look at some examples. We are "forced" to get up for a 7:30 a.m. class. He was "forced" off the road by another car. A push or pull is, perhaps, the simplest way to show force. Newton's three laws of motion provide a basis for discussing forces that act on some object. In this chapter you will be introduced to Newton's three laws of motion.

Remember as you read this chapter that the metric system of measurement is used in physics. The unit of length is the *meter*. One meter is equal to 3.28 feet. The *kilogram* is the unit of mass. One kilogram weighs 2.21 pounds at the earth's surface. Time is measured in *seconds*. This system is called the meter-kilogram-second, or MKS system.

Newton's First Law of Motion

First law. Newton's first law reads: "Every body continues in its state of rest, or of uniform motion, in a straight line unless it is compelled to change that state by forces impressed on it." *Uniform* here means constant. *Motion* refers to velocity. If the velocity is constant, both in magnitude and direction, the acceleration is zero. The velocity can be changed only by applying a force.

What this law suggests. When an object is seen to be accelerating, Newton's first law suggests looking for the force that causes the acceleration. Often this force is obvious. At other times it is not. A batted baseball is accelerated by the force that the bat exerts on the ball. The force is obvious. A ball dropped from the window accelerates toward the ground. What is the force that accelerates it? The force is not so obvious. The answer, however, is known to be gravity. The gravitational force between the earth and the ball accelerates it.

Newton's Second Law of Motion

What causes resistance? Suppose that you kicked different kinds of objects on the floor with the same force. Would all of the objects achieve the same acceleration? Surely they would not. Every object offers some resistance to being put into motion. Resistance is caused by the property of mass known as inertia. *Inertia* is the property or tendency of a body to resist any change in its state of motion, be it starting, stopping, or changing it from a straight-line path. The greater the weight, the greater the mass, the greater the inertia, the greater is the resistance to motion.

Assign a number to measure mass. A simple balance can be used to assign a number to measure mass. It compares the unknown mass with the known mass. See Figure 4-1. The units used are kilograms. The symbol for kilogram is kg.

Figure 4-1. **Measuring Mass**

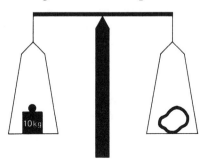

From: Joseph Priest, **Energy for a Technical Society**, ©1975, Addison-Wesley Pub. Co., Inc., Reading, Massachusettes, p. 31. Reprinted with permission.

Second law, force, and equations. Newton's second law quantifies these ideas discussed above. The second law states: "Acceleration experienced by an object is proportional to the net force acting on it and is inversely proportional to its mass." It then defines *force* as the amount of acceleration given a mass. In equation form you can say:

$$\text{Acceleration} = \frac{\text{net force}}{\text{mass}}$$

$$\mathbf{a = F/m} \quad \text{or} \quad \mathbf{F = ma}$$

The smaller the mass of an object acted upon by a given force, the greater its acceleration and hence, final velocity. See Figure 4-2.

Figure 4-2. **Mass and Acceleration**

When the same force is applied to bodies of different masses, the resulting accelerations are inversely proportional to the masses. Succesive positions of blocks of mass, **m, 2m,** and **3m,** are shown at 1-s intervals while identical forces of **F** are applied.

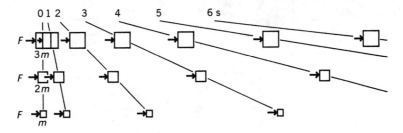

From: Arthur Beiser, **Physics,** 2nd Edition (Menlo Park: CA: The Benjamin/Cummings Pub. Co., 1978), p. 67. Reprinted by permission of the publisher.

The Newton. It is convenient to use a special unit for force. In the SI system of units (Systeme International Version of the Metric System), the unit used for force is the newton. Newton is abbreviated: *N.* A *newton (N)* is that force which, when applied to a 1 kg mass, gives it an acceleration of 1 m/s^2. Here, as in Figure 4-2, "s" refers to seconds and "m" refers to mass.

Newton's Third Law of Motion

Third law. Newton's third law states: "If one object (call it A) exerts a force on another object (call it B), then B exerts an equal but oppositely directed force on A." For example, suppose a moving car exerts a force on (strikes) an auto initially at rest. The struck auto is accelerated. It is pushed forward. The struck auto, at the same time, exerts an equal but oppositely directed force on the moving auto. This force tends to push the moving auto backwards. See Figure 4-3.

Figure 4-3. **Forces on Two Colliding Cars**

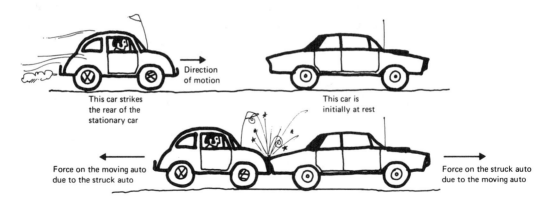

From: Joseph Priest, **Energy for a Technical Society**, ©1975, Addison-Wesley Pub. Co., Inc., Reading, Massachusettes, p. 32. Reprinted with permission.

Usefulness of the third law. Newton's third law is useful when trying to understand how things like atoms, molecules, and earth-satellite systems are held together. It is also useful when trying to understand the function of a jet engine.

Summary

Newton's first law of motion tells us that without a net force acting on it, a body at rest will remain at rest, and a body in motion will continue in motion at a constant velocity.

The second law of motion tells us that a net force acting on a body causes it to have an acceleration proportional to the magnitude of the force and inversely proportional to the body's mass. The acceleration is the same direction as the force.

The third law of motion tells us that when a body exerts a force on another body, the second body exerts a force on the first body too. This force is of the same magnitude, but it is in the opposite direction.

Exercises

1. Suppose that you had two identically shaped footballs, one filled with air, the other with sand. What difference in motion would you expect when they are kicked from a resting position on the ground with the same force?

2. Alvin Toffler, author of **Future Shock**, writes: "The acceleration of change in our time is, itself, an elemental force." What comparison can you make between this statement and Newton's second law of motion?

3. Can we conclude from the third law of motion that a single force cannot act upon a body?

Suggestions for Further Reading

Hewitt, Paul G. **Conceptual Physics**. Boston: Little, Brown, 1974.

Blatt, Frank J. **Principles of Physics**, Boston: Allyn and Bacon, Inc., 1983.

Chapter 7
Work

After reading this chapter, you will be able to:

> describe work from the physics standpoint and write the
>   mathematical equation for work
> describe what happens when work is done against gravity and
>   write the mathematical formula for it

In everyday language, the word "work" is associated with labor or toil. In this sense it can be thought of as "a physical or mental effort directed toward the production of something." In this chapter you will learn about work from the physics standpoint.

Work from the Physics Standpoint

Work and its equation. From the physics standpoint, physical effort directed toward the production of something is supplied by a force. Work is said to be done when a force acts on a mass while it moves through some distance. A physical quantity called *work* can be defined as follows:

> The work done by a force acting on an object which moves in the same direction as the force is equal to the magnitude of the force multipled by the distance through which the force acts.

> The definition of work in equation form reads:

$$\text{Work} = \text{force} \times \text{distance}$$
$$\mathbf{W} = \mathbf{Fs}$$

Measuring work. Work can be measured in units of foot pounds and joules. The foot pound (ft. lb.) is the unit of work in the British system of units. One *foot pound* is equal to the work done by a force of one pound acting through a distance of one foot.

For example, a man may push a refrigerator. The amount of work he does is equal to the magnitude or amount of force he applies times the distance he moves the refrigerator. He may push the refrigerator 10 feet with a force of 20 pounds. If he does this, he does 20 lb. x 10 ft. = 200 foot pounds of work.

The joule (J) is the unit of work in a more recent system of units, the Systeme International (SI), the current version of the metric system. One *joule* is equal to the work done by a force of 1N (newton) acting through a distance of 1m(meter), i.e. 1J = 1N x m. A *newton* is a metric unit of force that is equivalent to a force of about 1/4 pound.

It's important to realize, however, that a force may be exerted, yet no work may be done. This happens if the force does not act through a distance. For example, a man pushing against a brick wall does no work on the wall if the wall does not move. The product of force and distance is zero because the distance is zero. Only a force that gives rise to motion is doing work.

Positive work and negative work. It is important to recognize that a force can act in one of two ways. First, it can act in the same direction as the mass is moving (positive work). Second, force can act opposite to the direction in which the mass is moving (negative work). In both cases work is being done.
(See Figure 7-1, parts a and b.)

Figure 7-1. **Positive Work and Negative Work**

(a)   Force acts in the same direction that the car moves. This tends to increase the car's speed. This is a case of positive work.

(b)   Force acts in a direction opposite to that in which the car moves. This tends to decrease its speed. This is a case of negative work.

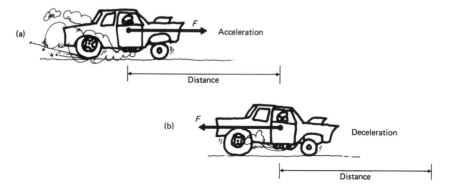

From: Joseph Priest, **Energy for a Technical Society,** ©1975, Addison-Wesley Pub. Co., Inc., Reading, Massachusettes, p. 34. Reprinted with permission.

Net work. The *net work* on an object is the algebraic sum of the positive and negative works done by each force acting on the object. Suppose that two forces of +100 N and -50N caused an object to be moved a distance of 2 meters. Then the work done to the two forces would be:

$$(+100) \ (+2) \ = \ +200J$$
$$\text{and}$$
$$(-50) \ (+2) \ = \ -100J$$

The net work would be +200 - 100 = +100J

Work Done Against Gravity

Acceleration due to gravity. The amount of work done lifting an object against gravity can be computed. It is necessary first to explain acceleration due to gravity and weight. When gravity is the only force acting, it turns out that the acceleration is the same for all masses. This acceleration is given the symbol "g." It is numerically equal to 9.8 meters per second per second. That is, the velocity changes by 9.8 meters per second for every second that the object falls. Often meters per second per second is abbreviated m/sec^2.

Weight and its symbols. The force of gravity on an object is the same as its weight. The *weight* of an object is simply its mass multiplied by the acceleration due to gravity. In symbols: **w**(weight) + **m**(mass) x **g** (acceleration due to gravity), or **w = mg**.

Computing work done against gravity. It is easy then to compute the work done lifting an object against gravity. Keep two points in mind. First, the force of gravity on an object of mass, **m**, is the same as its weight, **w = mg**. Second, in order to raise the object to a height, **h,** above its original position, a force of **mg** must be exerted on it.

To compute work done against gravity then, you use the following equation:

$$\mathbf{W \ = \ Fs}$$
$$\mathbf{= \ mgh}$$

Lifting an object of mass, **m**, to a height, **h**, requires the performance of the amount of work, **mgh.** See Figure 7-2.

Figure 7-2. **Gravity and Work**

(a)   The work **mgh** must be done to lift an object of
       mass **m** to a height **h.**
(b)   When an object of mass **m** falls from a height **h,**
       the force of gravity does the work **mgh** on it.
(c)   The force of gravity does no work on objects
       that move parallel to the earth's surface.

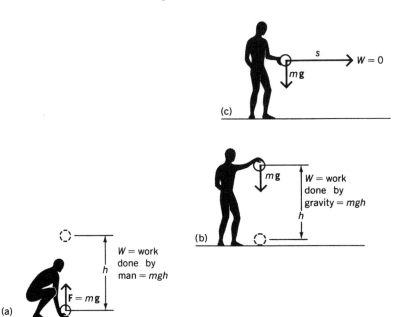

From: Arthur Beiser, **Physics,** 2nd Edition (Menlo Park: CA:
The Benjamin/Cummings Pub. Co., 1978), p. 146. Reprinted by
permission of the publisher.

Importance of route taken. It is important to note that only the height **h** is involved in work
done against the force of gravity. The particular route taken by an object being raised is
not significant. See Figure 7-3.

Figure 7-3. **The Importance of the Route Taken**
In the absence of friction, the work done in lifting a
mass **m** to a height **h** is **mgh** regardless of the exact
path taken.

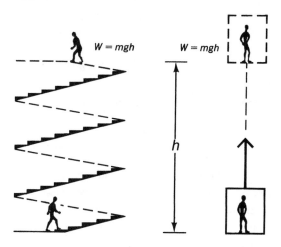

From: Arthur Beiser, **Physics**, 2nd Edition (Menlo Park: CA:
The Benjamin/Cummings Pub. Co., 1978), p. 147. Reprinted by
permission of the publisher.

Excluding any frictional effects, exactly as much work must be done to climb a flight of
stairs as to go up in an elevator to the same floor (though not by the person involved).

Summary

Work is force acting through a distance. The equation for work is $W = Fs$. The force must
give rise to motion to produce work. In the SI system, the unit of work is the joule. In the
British system, it is the foot-pound. A force can act in the same or opposite direction in
which the mass is moving. Net work is the algebraic sum of the positive and negative works
done by each force.

It's easy to compute work done against gravity. Essentially:

$$F = mg \text{ and } s = h,$$

so,

$$W = Fs$$
$$= mgh$$

The route taken by the object being raised is not important.

Exercises

1. A person strongly exerts himself in attempting to push a stalled auto. Nevertheless, the auto does not move. Is work done in the scientific sense? Is energy expended?

2. Work must be done on a car to reduce its speed from 50 to 40 miles per hour. Is this work positive or negative?

3. How much work is done in lifting a 40 kg load of building material to a height of 30 meters in a building under construction?

4. A person applies a 200 N vertical force to lift a box through a distance of 1 meter. How many joules of work are done?

Chapter 3
Atoms and Molecules

The atomic theory was developed only a little more than 150 years ago. Two and a half centuries ago, however, Robert Boyle and others developed the idea that the many different varieties of matter are composed of only a few chemical elements. This concept led to the idea that there are only as many different kinds of atoms as there are chemical elements. In this form, the atomic theory was utilized by John Dalton, an English scientist. He used this theory to explain the laws of chemical combination that were formulated in the 1800s.

In this chapter you will learn about Dalton's assumptions regarding atoms and molecules and the Law of Conservation of Mass. Additional information is also included about molecules.

Dalton's Assumptions

Assumptions. Dalton made two assumptions about atoms and molecules. First, Dalton hypothesized that all the atoms of a particular element are alike and have the same mass. An *atom* is the smallest particle of an element. An *element* is a substance composed of only one type of element; it cannot be broken into other substances by chemical means. Second, Dalton assumed that chemical combination involves chemical bonding. This bonding of atoms of each of the combining elements forms molecules of the resulting compound. Bonding occurs in a predictable manner.

Law of Conservation of Mass. Dalton's atomic theory helped explain the Law of Conservation of Mass. The Law of Conservation of Mass tells us that there is no detectable gain or loss of mass during any physical or chemical change. Note that the Law of Conservation of Mass is applied to chemical reactions as well as to physical transformations. Atoms involved in both kinds of processes keep their identities. In a chemical reaction, the molecules of the products contain the same atoms that were initially present in the molecules of the reactants. When charcoal burns, for example, there are two reactants. One is an atom of carbon. The second is a molecule of oxygen. One atom of carbon combines with one molecule of oxygen. The product or result is a carbon dioxide molecule. Each individual atom retains its original mass. The sum of masses of the molecules of the product is the sum of the masses of the molecules of the reactants.

Molecules

Molecule defined. From the standpoint of chemistry, a *molecule* may be thought of as the smallest particle of a substance as it normally exists. Three additional points about molecules may be specified. First, a molecule can be described as a minute, uncharged particle of a substance. Second, it moves about as a whole. Third, its component parts do not become detached during motion.

Molecules of compounds and elements. A molecule of a compound consists of more than one kind of atom. See Figure 3-1.

Figure 3-1. **Molecules of Compounds**

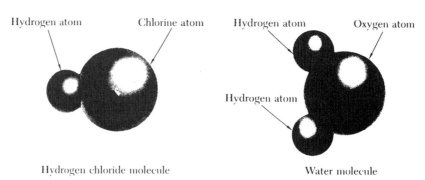

Hydrogen atom   Chlorine atom   Hydrogen atom   Oxygen atom

Hydrogen atom

Hydrogen chloride molecule   Water molecule

From: Lawrence P. Eblin, **The Elements of Chemistry**, 2nd Edition (Orlando: Harcourt Brace Jovanovich, 1970), p. 13. Reprinted by permission of the publisher.

A molecule of an element, however, usually consists of two or more atoms of the same kind. See Figure 3-2.

Figure 3-2. **Molecule of an Element**

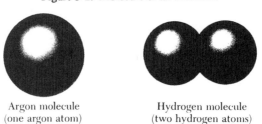

Argon molecule   Hydrogen molecule
(one argon atom)   (two hydrogen atoms)

From: Lawrence P. Eblin, **The Elements of Chemistry**, 2nd Edition (Orlando: Harcourt Brace Jovanovich, 1970), p. 13. Reprinted by permission of the publisher.

Summary

Dalton's assumptions about atoms and molecules helped explain the Law of Conservation of Mass. This law tells us that there is no detectable gain or loss of mass during any physical or chemical change. The sum of masses of the molecules of the product is the sum of the masses of the molecules of the reactants. A molecule of a compound consists of more than one kind of atom. A molecule of an element usually consists of two or more atoms of the same kind.

Exercises

1. What is an atom?
2. Define molecule.
3. Describe how molecules of compounds are formed through chemical combination.
4. What is the Law of Conservation of Mass?

Suggestion for Further Reading

Davis, Raymond E., Gailey, Kenneth D. and Whitten, Kenneth W.
Principles of Chemistry. Philadelphia: Saunders College Publishing, 1984.

Chapter 5
Periodic Table of Elements

What, exactly, is the periodic table?

What elements are included in the horizontal rows?

What groups are included in the vertical columns and what are their names?

The *periodic table* is an array of elements arranged in order of their atomic numbers. They are arranged in 7 rows and 18 columns. Elements in any column have similar chemical properties. This chapter introduces you to the periodic table.

Rows

Horizontal rows. The horizontal rows in the table are chemical periods. The elements in any one row make up a series or period. There are certain points to note about the rows. First, there are seven rows, numbered one through seven. Second, the number of elements in each row or period (1-7) increases in the order 2, 8, 8, 18, 18, 32, 17... (the last period is incomplete). Third, most modern tables show 15 elements from each of two long periods printed below the table. See Table 5-1.

Last element. The last element of a row or period is always a noble gas. Noble gases are nonatomic, or single atom gases. They have very little chemical reactivity. Examples of noble gases are helium, neon, and argon.

Groups

Vertical columns. The vertical columns in the table are chemical groups. The elements in any one column therefore make up a group.

Elements in a group or column. Elements in a group are a family of elements having similar properties. They're more like each other than they're like elements of other groups.

Table 5-1. **The Periodic Table of Elements**

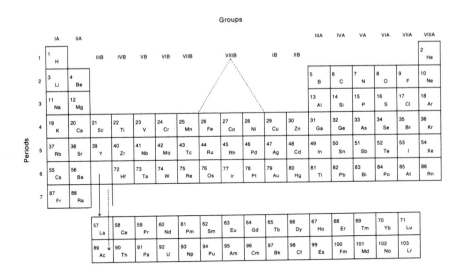

From: Richard M. McCurdy, **Qualities and Quantities: Preparation for College Chemistry** (Orlando: Harcourt Brace Jovanovich, 1975), p. 106. Reprented by permission of the publisher.

Additional points can be made about elements in the groups which are designated by Roman numerals in two series, A and B. First, the physical and chemical properties of elements in a group quite regularly change in qualities with increasing atomic number. For example, the melting points (Celsius) of the alkali metals are $_3$Li, 180°; $_{11}$Na, 98°; $_{19}$K, 64 °; $_{37}$Rb, 39°; $_{55}$Cs, 29°. See Figure 5-2.

Figure 5-2. **Melting Points of the Alkali Metals**

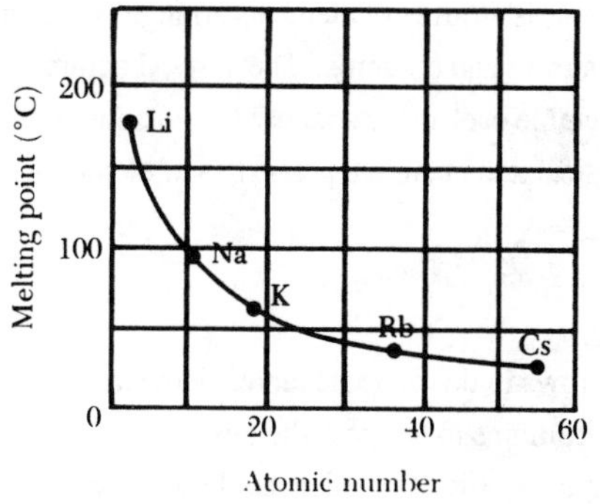

From: Lawrence P. Eblin, **The Elements of Chemistry**, 2nd
Edition (Orlando: Harcourt Brace Jovanovich, 1970), p. 67.
Reprinted by permission of the publisher.

Second, elements in the same group form compounds with similar formulas. For example, formulas for oxides of Group VB elements are: V_2O_5, Nb_2O_5, Ta_2O_5. Third, in the A groups, metallic properties are accentuated with an increase in the atomic number. Fourth, in the A group, nonmetallic properties diminish as the atomic number increases. Fifth, the first member of a group is often more different from the other members of the group than the other members are different from each other. For example, fluorine does not resemble chlorine as much as chlorine resembles bromine. Also boron is quite different from the other elements of Group IIIA. It is the only nonmetal in the group. Sixth, the members of an A group somewhat resemble members of the corresponding B group. For example, consider manganese in Group VIIB and chlorine in Group VIIA. They both form acids with similar formulas and properties.

Group names. Certain groups have been given names. Group IA metals (excluding hydrogen) are called alkali metals. Group IIA metals are the alkaline earth metals. On the other side of the table, Group VIIA are the halogens. Group VIIIA are the noble gases.

What location tells us. Where an element is located in the table tells us its physical and chemical properties. The elements on the left side, for example, are strongly metallic. The elements on the right side are nonmetallic. Then, some elements are between the metals and nonmetals. They are sometimes called metalloids. For example, consider arsenic and selenium. They're sold under the labels, "arsenic metal" and "selenium metal." Chemists would agree, however, that these elements should be classified as nonmetals. Remember also that what is true about one member of a group is usually true about the others in the same group.

Summary

The periodic table is an array of elements arranged in order of their atomic numbers. They are arranged in rows and columns. The vertical columns are chemical groups. Elements in a group resemble each other somewhat. Position within the table indicates something about the physical and chemical properties of elements too.

Exercises

1. What are rows in the periodic table? How many are there?
2. What are columns in the periodic table?
3. Name some specific things that can be said about the groups of elements.
4. What does position within the table tell you about the elements?
5. Construct a blank periodic table and indicate location of the following: sets of elements, alkali metals, halogens, noble gases.

Chapter 7
Relationship of Volume and Pressure of Gases

After reading this chapter you will be able to:

    explain how volume and pressure of gases are related; and

    compute final volume when volume changes because of

    pressure changes.

Long ago, experiments were conducted to determine how the volume and pressure of gases are related. In this chapter, you will learn about this relationship. You will also learn how to compute final volume when volume changes due to pressure changes.

Early Experimentation

Experiment. Sir Robert Boyle tried to discover if the volume and pressure of a gas were related. He used an apparatus with gas and mercury to measure the relationship. See Figure 7-1.

Figure 7-1. **Apparatus for Determining Relationship**
of Volume and Pressure of Gases

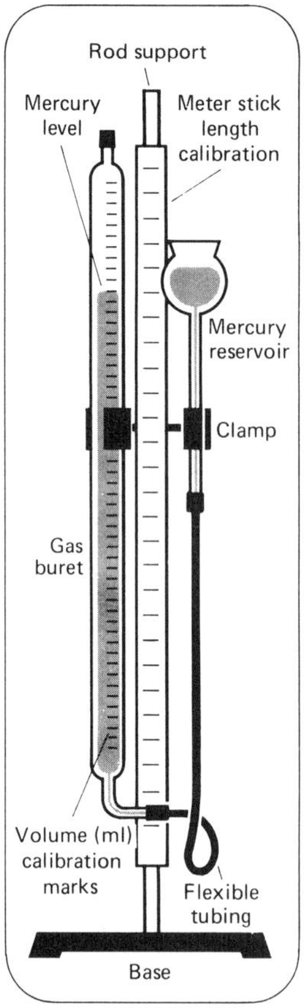

From: Richard M. McGurdy, **Qualities and Quantities:**
Preparation for College Chemistry (Orlando: Harcourt
Brace Jovanovich, 1975), p. 65. Reprinted by permission
of the publisher.

A sample of gas (e.g., air, nitrogen, or carbon dioxide) was trapped in the apparatus. It was trapped over mercury in a closed device used to measure volume. The mercury acted as a piston to change the volume of the gas when the mercury reservoir was raised or lowered. The surface level of the mercury was related to the amount of pressure exerted.

Result. Boyle found that for gases, as the pressure increased the volume was decreased. Volume was decreased in proportion to the amount of pressure increase. Using a constant amount of gas at a constant temperature, he was able to show this relationship. Look at the figures in Table 7-1. Notice the definition of torr in Table 7-1 also.

Table 7-1. **Pressure versus Volume (at constant
temperature)**

Pressure (torr)	Volume (ml)	PV Product
950	30.3	2.88×10^4
750	38.4	2.88×10^4
500	57.6	2.88×10^4
350	82.3	2.88×10^4
300	96.0	2.88×10^4

A *torr* is the pressure exerted by a column of mercury one millimeter high.

From: Richard M. McGurdy, **Qualities and Quantities: Preparation for College Chemistry** (Orlando: Harcourt Brace
Jovanovich, 1975), p. 65. Reprinted by permission of the
publisher.

These figures can be placed on a graph. The curve in the graph shows an inverse relationship.
As pressure increases, volume decreases. Look at the solid line in Figure 7-2.

Figure 7-2. **Pressure and Volume of Gases**

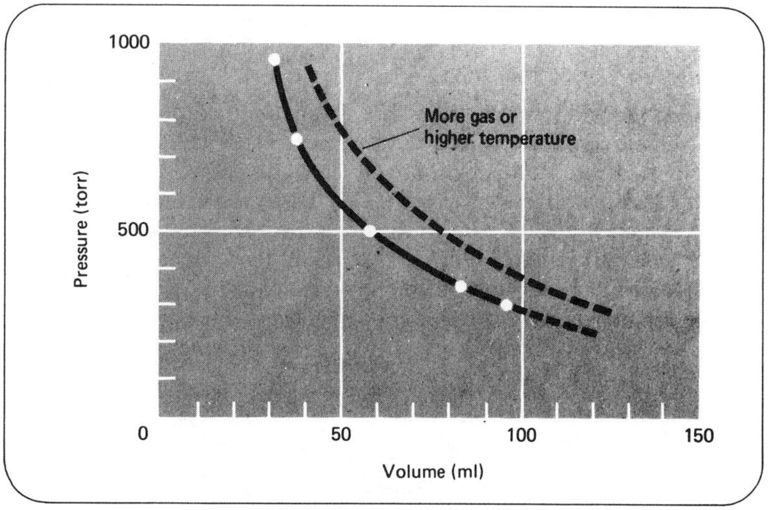

From: Richard M. McGurdy, **Qualities and Quantities: Preparation for College Chemistry** (Orlando: Harcourt Brace Jovanovich,
1975), p. 66. Reprinted by permission of the publisher.

Equation for relationship. The equation for the relationship between volume and pressure is $V = k/P$. V is volume. P is pressure. k is a proportionality constant. This constant, k, depends on the temperature and the amount of gas. As the pressure increases, the volume decreases. The fraction k/P gets smaller. It never becomes zero or negative, however. As the pressure decreases the volume increases. Volume becomes very great as the pressure nears zero.

A sample with more gas or higher temperature could be used too. It would result in a similar curve (dashed line) further from the axis. See Figure 7-2.

Conclusions

How volume changes. The experiment shows that at constant temperature, the volume of the same amount of gas changes with pressure. It changes in a specific way: the product of the new pressure and volume always equals the product of the initial pressure and volume. Look at the following equation:

$$P_f V_f = P_i V_i$$

The *subscript f* means final. The *subscript i* means initial.

Figuring final volume. Knowing the initial volume then allows you to figure the final volume. To figure final volume, you multiply the initial volume by the ratio of pressures:

$$V_f = V_i \times \frac{P_i}{P_f}$$

Sample problem and solution. For this problem, suppose you have a given sample of a gas. Two facts are provided. The gas is at room temperature. It has a volume of 35.0 milliliters at a pressure of 725 torr. The question about this sample of gas is: "What will be the volume of this sample of gas at 760 torr?"

To find the solution, follow two steps. First, write out the equation $P_f V_f = P_i V_i$ using real figures. The product of 35.0 ml and 725 torr is a constant at a fixed temperature and amount of gas. The product of the new volume and 760 torr is the same constant. Therefore you will write: 35.0 ml x 725 torr = $V_f$ml x 760 torr. Two things equal to the same thing are equal to each other. Second, use real figures in the formula figuring final volume:

$$V_f = V_i \text{ ml } (35.0 \text{ ml}) \times \frac{P_i \,(750 \text{ torr})}{P_f \,(760 \text{ torr})} = 33.4 \text{ ml}$$

Therefore, 33.4 ml is the final volume.

Common Sense

Common sense solution. Actually, volume changes caused by pressure changes can be figured by "common sense." Two points must be realized. First, an increase of pressure squeezes a gas; the final volume will be the initial volume times a ratio of pressures less than one. Second, a decrease of pressure allows a gas to "press out" or expand; the final volume will be the initial volume times a factor greater than one.

Two sample problems and solutions using common sense. For the first problem, suppose the pressure on 10 liters of gas is doubled. The question for this problem is: "What is the new volume?" Remember, increasing the pressure squeezes a gas. To reach the solution, multiply the initial volume by a pressure ratio smaller than one:

$$V_f = 10 \text{ liters} \times \frac{1}{2} = 5 \text{ liters}$$

For the second problem, suppose the initial pressure on 10 liters of gas is 760 torr. The question for this problem is: "What is the volume when the pressure is decreased to 380 torr?" Remember, decreasing the pressure allows a gas to expand. To reach the solution, multiply the initial volume by a factor greater than one:

$$V_f = 10 \text{ liters} \times \frac{760}{380} = 20 \text{ liters}$$

Summary

At constant temperature, the volume of a constant amount of gas will change with pressure. An increase of pressure squeezes a gas. A decrease of pressure allows a gas to expand. The final volume can be computed from the initial volume.

Exercises

1. The initial pressure on 20 milliliters of gas is 600 torr. What would the volume of gas be at 650 torr?
2. The pressure on 15 liters of gas is tripled. What will the new volume be?
3. The initial pressure on 30 liters of gas is 750 torr. What is the volume when the pressure is decreased to 400 torr?

Answer Key
For Studying the Sciences

Skill 1, Exercise 1
1. **Understanding Biology**
2. No
3. Burton S. Guttman and Johns W. Hopkins III
4. Harcourt Brace Jovanovich, Inc.

Skill 1, Exercise 2
5. 1983
6. The Benjamin/Cummings Publishing Company, Inc.

Skill 1, Exercise 3
7. Life at the community level
8. page 91
10. Biology at the population level
11. The growth and regulation of populations

Skill 1, Exercise 4
12. To provide a comprehensive introduction to the nature and functioning of living systems within the biosphere on earth.
13. Yes

Skill 1, Exercise 5
14. A microorganism that requires oxygen for respiration
15. One of the air sacs in the lungs

Skill 1, Exercise 6
16. pages 444-445
17. pages 292-293
18. Pyruvic acid
19. Homeostasis
20. 3

Skill 1, Exercise 7
21. **Foundations of College Chemistry**
22. No
23. Morris Hein

24. 5th edition
25. Brooks/Cole Publishing Company

Skill 1, Exercise 8
26. 1980
27. John Wiley and Sons, Inc.

Skill 1, Exercise 9
28. Atoms in Combination
29. page 749
30. 8
31. Useful Mathematics
32. Powers of Ten and Logarithms

Skill 1, Exercise 10
33. elementary algebra and simple trigonometry
34. To provide a basic foundation in physics for students who will need some competence in physics in their careers.

Skill 1, Exercise 11
35. In a *covalent bond* between adjacent atoms of a molecule or solid, the atoms share one or more electron pairs.
36. The *Doppler effect* refers to the change in frequency of a wave when there is relative motion between its source and an observer.

Skill 1, Exercise 12
37. pages 490-91
38. pages 403-410
39. Electric charge
40. Nuclear power
41. 3

Skill 1, Exercise 13
42. **Physics**
43. Yes, for Scientists and Engineers
44. Raymond A. Serway

45.   Saunders College Publishing

Skill 1, Exercise 14

46.   Fundamental Ideas of Chemistry
47.   page 4
48.   3
49.   Stoichiometry
50.   The Mole

Skill 1, Exercise 15

51.   pages 378-380

52.   pages 222-227
53.   Barbiturates
54.   10

Skill 1, Exercise 16

55.   One of the reproductive cells of
      a sexually reproducing organism.
56.   A representation of the structure of
      a genome as determined by
      mapping experiments.

Skill 2, Exercise 1

1.   Algae
2   a and c
3.   This chapter is about the nature of algae in general and the
     specific characteristics of blue-green algae.
4.   Chapter 12: Algae
     What Are Algae Like?
          Four general traits
          Importance to aquatic communities
     Blue-Green Algae
          Basis for color
          Cell structure
          How blue-green algae reproduce
          Habitat
5.   They contain blue pigment, which combines with green chlorophyll
     to produce a blue-green color.
6.   in ponds, streams, and moist places on land.
7.   a, c
8.   The major purpose for reading the chapter would be to learn the four
     traits shared by all algae and the importance of algae to aquatic communities.
     One would also read to learn basic facts/information about blue-green algae,
     the basis for their color and cell structure, their method of reproduction, and
     their typical habitat.

Skill 2, Exercise 2

9.   Protozoa
10. b and c
11. This chapter describes protozoa in general. In addition, it discusses two
     common types of protozoa—amoebae and ciliates.

12.   Chapter 13: Protozoa
     Protozoa in General
     Amoebae
        Forming pseudopodia
        Food-getting structures
        Habitat
     Ciliates
        General description
        Paramecium

13.   one-celled, nongreen, animal-like protists

14.   getting food and moving about

15.   b and c

16.   The major purpose for reading the chapter would be to learn the characteristics shared by all protozoa and more specifically to learn the unique features of two common types of protozoa, amoebae and ciliates. One would read to learn how amoebae form pseudopodia, how the latter enable them to move about and get food, and where amoebae live, as well as to learn the characteristics of ciliates including those of paramecium, a familiar ciliate.

Skill 2, Exercise 3

17.   Origin of Genetics

18.   a and c

19.   This chapter is about Mendel's experiments and how they led him to the law of dominance.

20.   Chapter 11: Origin of Genetics
     Mendel's Initial Experiment
        Purpose
        Plants used in the experiment
        Mendel's accomplishment
        Results
        Law of dominance
     Further Experimentation
        F_2 generation
        Results

21.   pure plants having one trait and pure plants with the opposite trait

22.   For each trait there is one form which "dominates" the other (opposite) form.

23.   b

24.   The major purpose for reading the chapter would be to learn why Mendel conducted his first experiment with genetics, what kind of plants were used in this experiment, what was accomplished and the results, and how these results led to the Law of Dominance. One would also read to learn how Mendel

conducted further experimentation yielding the F_2 generation and the specific
results of this experimentation.

Skill 2, Exercise 4

25.   Speed, Velocity, and Acceleration
26.   b and c
27.   This chapter discusses the basic theories of speed, velocity, and acceleration
      and provides mathematical formulas for computing each.
28.   Chapter 2: Speed, Velocity, and Acceleration
      Speed
          Speed and its formula
          Distance
      Velocity
      Acceleration
          Acceleration response
          Direction and acceleration
29.   the rate at which distance is being covered by the body
30.   a specification of both its speed and the direction in which it is moving
31.   b and c
32.   The major purpose for reading the chapter would be to learn the fundamental
      aspects involved in the theories of speed, velocity, and acceleration and how to
      write the mathematical formulas for computing them.

Skill 2, Exercise 5

33.   Newton's Three Laws of Motion
34.   c
35.   This chapter discusses Newton's three laws of motion, which provide
      a basis for discussing forces that act on an object.
36.   Chapter 4: Newton's Three Laws of Motion
      Newton's First Law of Motion
          First law
          What this law suggests
      Newton's Second Law of Motion
          What causes resistance?
          Assign a number to measure mass
          Second law, force, and equations
          The Newton
      Newton's Third Law of Motion
          Third Law
          Usefulness of the third law
37.   Without a net force acting on it, a body at rest will remain at rest, and a
      body in motion will continue in motion at a constant velocity.

38. When a body exerts a force on another body, the second body exerts a force on the first body too. The second force is of the same magnitude, but it is in the opposite direction.

39. c

40. The major purpose for reading the chapter would be to learn what is involved in Newton's three laws of motion and what these suggest. With regard to the second law of motion, one would read to learn, more specifically, what causes resistance, how mass is measured, what formula expresses this law, what the Newton is, and one would read to learn how the third law is useful.

Skill 2, Exercise 6

41. Work

42. b and c

43. This chapter discussed "work" from the physics standpoint including its mathematical equation. It also discusses what happens when work is done against gravity and the mathematical formula for calculating it.

44.   Chapter 7: Work
Work from the Physics Standpoint
      Work and its equation
      Measuring work
      Positive work and negative work
      Net work
Work Done Against Gravity
      Acceleration due to gravity
      Weight and its symbols
      Computing work done against gravity
      Importance of route taken

45. $W = Fs$

46. the joule

47. a and d

48. The major purpose for reading the chapter would be to learn how to describe/define work from the physics standpoint, write its equation and measure it, differentiate between positive and negative work, and define/describe net work. One would also read to learn what happens when work is done against gravity and how to compute it, how to explain acceleration due to gravity, how weight is defined and the symbols used in its equation, and what effect the route taken by the object has on the work done against gravity.

Skill 2, Exercise 7

49. Atoms and Molecules

50. c

51.   This chapter is about Dalton's assumptions regarding atoms and molecules and the Law of Conservation of Mass. Additional information is included about molecules.

52.   Chapter 3: Atoms and Molecules
      Dalton's Assumptions
          Assumptions
          Law of Conservation of Mass
      Molecules
          Molecule defined
          Molecules of compounds and elements

53.   There is no detectable gain or loss of mass during any physical or chemical change.

54.   two or more atoms of the same kind

55.   a and c

56.   The major purpose for reading this chapter would be to learn what assumptions were made by Dalton, what the Law of Conservation of Mass tells us, how to define molecule, and how molecules of compounds differ from those of elements.

Skill 2, Exercise 8

57.   Periodic Table of Elements

58.   a and c

59.   This chapter is about the periodic table, the elements included in its horizontal rows, and the groups included in its vertical columns.

60.   Chapter 5: Periodic Table of Elements
      Rows
          Horizontal rows
          Last element
      Groups
          Vertical columns
          Elements in a group or column
          Group names
          What location tells us

61.   an array of elements arranged in order of their atomic numbers

62.   chemical groups

63.   b and d

64.   The major purpose for reading this chapter would be to learn of what the periodic table consists including what elements are in its horizontal rows, what the last element in a row is, what groups of elements (and their names) are found in its vertical columns and what location in the table tells us about the elements.

Skill 2, Exercise 9

65.   Relationship of Volume and Pressure of Gases.

66.   b and c

67.   This chapter discusses the relationship of volume and pressure of gases. How to compute final volume when volume changes due to pressure changes is also explained.

68.   Chapter 7: Relationship of Volume and Pressure of Gases

      Early Experimentation

         Experiment

         Result

         Equation for relationship

      Conclusions

         How volume changes

         Figuring final volume

         Sample problem and solution

      Common Sense

         Common sense solution

         Two sample problems and solutions using common sense

69.   Gas is squeezed; it is reduced in volume.

70.   Gas expands; it increases in volume.

71.   a and c

72.   The major purpose for reading this chapter would be to learn what early experimentation told us about the relationship of volume and pressure of gases, what equation is used to express this relationship, how to compute final volume when volume changes due to variation in pressure, and how to use common sense when figuring volume changes caused by pressure changes.

Skill 3, Exercise 1

1.   a.   Digestion

      b.   process of breaking large molecules into smaller ones by chemical and physical means

      c.   Ask your instructor to read your answer.

2.   a.   Pasteurization

      b.   process of heating and then rapidly cooling a substance to kill dangerous bacteria

      c.   Ask your instructor to read your answer.

3.   a.   Evaporation

      b.   process whereby a liquid becomes a gas and likewise changes for the liquid to the gaseous state

      c.   Ask your instructor to read your answer.

4. a. Oxidation
    b. the combination of a substance with oxygen which may occur at ordinary tempera-
        tures involving some substances such as yellow phosphorous or oily rags; these
        substances react at a faster rate as temperature increases
    c. Ask your instructor to read your answer.
5. a. Kindling temperature
    b. the temperature at which a combustible substance bursts into flame
    c. Ask your instructor to read your answer.

Skill 3, Exercise 2

6. a. Chain reaction
    b. reaction between gaseous hydrogen and gaseous chlorine in sunlight
    c. Ask your instructor to read your answer.
7. a. Solute
    b. salt which is dissolved in a beaker of water
    c. Ask your instructor to read your answer.
8. a. Solvent
    b. water
    c. Ask your instructor to read your answer.
9. a. Viscosity
    b. viscosity of heavy lubricating oils is high especially in comparison to liquids such
        as ether or benzene
    c. Ask your instructor to read your answer.
10. a. Sublimation
    b. "dry ice" (solid carbon dioxide) evaporates directly to gaseous carbon dioxide at
        temperatures above -78.5°C and does not pass through the liquid state
    c. Ask your instructor to read your answer.

Skill 3, Exercise 3

11. a. Topsoil
    b. soil that contained little organic material and other substances and that did not hold
        water and air well
    c. Ask your instructor to read your answer.
12. a. Habit
    b. convince himself prior to each meal to eat only foods high in nutritional value
    c. Ask your instructor to read your answer.
13. a. Extinct
    b. an existing form of life
    c. Ask your instructor to read your answer.
14. a. Desert
    b. rainfall of more than 25 cm per year and vegetation that was continuous or nar-
        rowly spaced
    c. Ask your instructor to read your answer.

15. a. Acid
    b. in solution, it had a larger number of hydroxide ions than hydrogen ions
    c. Ask your instructor to read your answer.

Skill 3, Exercise 4

16. a. Adapted
    b. Three major types of adaptation are morphological, physiological and behavioral.
    c. Ask your instructor to read your answer.

17. a. Embryo
    b. Embryos are alive and have the same basic functions as any other living organism in that they secure food, obtain oxygen, rid themselves of wastes, and respond to their environment.
    c. Ask your instructor to read your answer.

18. a. External fertilization
    b. External fertilization occurs in sponges, jellyfish, most worms, fish and frogs.
    c. Ask your instructor to read your answer.

19. a. Sexual reproduction
    b. One set of DNA comes from each parent.
    c. Ask your instructor to read your answer.

20. a. Asexual reproduction
    b. In vegetative propagation, a new organism is produced from a nonsexual (vegetative) part of the parent organism, as through fission in one-celled organisms including bacteria.
    c. Ask your instructor to read your answer.

Skill 3, Exercise 5

21. im
22. con
23. im
24. re
25. Ex
26. com
27. inter
28. de
29. dis
30. mono

Skill 3, Exercise 6

31. acting directly to produce an effect; acting or producing effectively with a minimum of waste, expense, or unnecessary effort
32. the science of living organisms and life processes, including the study of structure, functioning, growth, origin, evolution, and distribution of living organisms
33. a fold or folding back; copy or reproduction; act or process of duplicating or reproducing something

34. series of changes which result in the formation of specialized body parts; to develop into specialized organs. Used esp. of embryonic cells or tissues
35. act of producing a counterpart, image or copy; (Biol) the sexual or asexual process by which organisms generate others of the same kind
36. act of applying; a specific use to which something is put: the application of science to industry
37. materials in which electric charge flows relatively freely
38. a putting together of parts or elements to form a whole
39. a broad sequence or range of related qualities, ideas, or activities; (Phys) the distribution of a characteristic of a physical system or phenomenon
40. the act of giving or sending out (radiation, for example); something that is emitted; the substance discharged into the air, esp. by an internal combustion engine

Skill 3, Exercise 7
41. c
42. d
43. f
44. a
45. b
46. h
47. g
48. k
49. j
50. i

Skill 3, Exercise 8
51. collisions
52. impedance
53. connective
54. immigration
55. coordination
56. immunological
57. internal
58. comparative
59. corroborate
60. collaborate

Skill 3, Exercise 9

TERM	Dictionary RESPELLING	MEANING
61. presuppose	(prē - sə- pōz′)	to presume or suppose in advance; to require or involve necessary as an antecedent

Ask your instructor to read your sentence.

TERM	Dictionary RESPELLING	MEANING
62. evolution	(ĕv -ə - lōō´ shən)	a gradual process in which something changes into a different and usually more complex or better form; (Biol) the theory that groups of organisms, as species, may change with passage of time so that descendants differ morphologically and psychologically from their ancestors

Ask your instructor to read your sentence.

| 63. ecology | (ĭ kŏl´ə - jē) | the science of the relationship between organisms and their environment |

Ask your instructor to read your sentence.

| 64. subspecies | (sŭb´ spē - shēz) | a subdivision of a taxonomic species, usually based on geographical distribution |

Ask your instructor to read your sentence.

| 65. extrapolate | (ĭk - străp ə - lāt) | to estimate (a values(s) of a function) for values of the argument not used in the process of estimation; infer from known values; to infer or estimate by extending or projecting known information |

Ask your instructor to read your sentence.

| 66. erode | (ĭ - rōd´) | to wear something away by or as if by abrasion: Waves erode the shore. to eat into or corrode |

Ask your instructor to read your sentence.

| 67. mutate | (myōō´ tāt) | to change the genetic material at particular locus sites along the chromosome |

Ask your instructor to read your sentence.

TERM	Dictionary RESPELLING	MEANING
68. transpose	(trăns - pōz′)	to reverse or transfer the order or place of something; to put into a different place or order; to alter form or nature

Ask your instructor to read your sentence.

69. monogamy	(mə - nŏg′ ə - mē)	the custom or condition of being married to only one person at a time; (Biol) one female and one male share a pair bond (on mate for life) and both parents typically care for their young

Ask your instructor to read your sentence.

70. dispersion	(dĭ - spûr′ zhən)	the state of being dispersed or broken up and scattered in various directions; (Phys) the separation of a complex wave into component parts according to a given characteristic such as frequency or wavelength; the separation of visible light into its color components by refraction or diffraction; (Chem) a suspension, such as smog or homogenized milk, of solid, liquid, or gaseous particles of colloidal size or larger, in a liquid, solid, or gasious medium

Ask your instructor to read your sentence.

Skill 4, Exercise 1

Blue-Green Algae

What?

ANS *Basis for color.* Blue-green algae contain blue pigment. In combination with chloro- phyll, which is green in color, the blue pigment gives algae their blue-green color. Some kinds of blue-green algae also contain a red pigment. Red pigment, in combina- tion with other pigments, creates a near black color in some "blue-green" algae.

What?

ANS

DEF *Cell structure.* Blue-green algae are organized in a simple manner. They are the simplest in organization of all algae in several ways. First, there isn't a definite nucleus in a single cell. As a result, DNA is scattered throughout the cell. *DNA is the material that carries the genetic information for reproduction.* [2] Second, chlorophyll in blue-green algae is usually attached to membranes. There are no chloroplasts to contain the chlorophyll. [3] Third, single cells are often arranged in chains or filaments as shown in Figure 12-1. [4] Fourth, there seems to be little division of labor among the cells. Each cell is like every other cell of the group. [5] Fifth, groups of cells are often enclosed within a protective jelly-like layer.

Figure 12-1. **Blue-Green Algae**

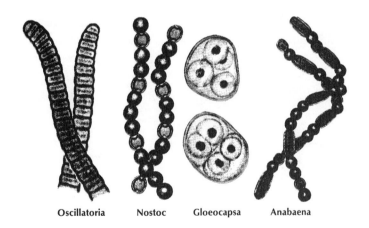

Oscillatoria   Nostoc   Gloeocapsa   Anabaena

From: Raymond F. Oram, Paul J. Hummer, Jr., and Robert C. Smoot, **Biology: Living Systems** (Columbus, Ohio: Charles Merrill Pub. Co., 1976), p. 314. Reprinted by permission of the publisher.

How?
ANS *How blue-green algae reproduce.* Blue-green algae reproduce asexually (without mating) through division of a single cell. New cells may be formed along a chain or filament through simple fission.

DEF *Fission* is reproduction in which a one-celled organism divides into two one-celled organisms. Occasionally, filaments are broken apart. Each section produces a new series of cells by fission. One kind of blue-green algae called Anabaena produces specialized cells that act as spores. These spores can develop into new filaments.

What?
ANS *Habitat,* Blue-green algae are commonly found in ponds, streams, and moist places on land. During hot summer months they multiply rapidly. As a result, public water supplies must be treated regularly during this season. Treatment is necessary to prevent rapid reproduction of blue-green algae and other microorganisms.

Skill 4, Exercise 2

Amoebae

How?
DEF *Forming pseudopodia.* One of the simplest protozoans is the amoeba. Amoebae are unique in how they move. To move, they develop pseudopodia. *Pseudopodia* are false feet formed by flexible
ANS plasma membranes and cytoplasm. The formation of pseudopodia involves certain stages. ¹Initially, the cytoplasm is in a fixed, gelantinous state. ²Then, chemical changes cause part of the cytoplasm to become a fluid. This fluid, which is like a true solution, is known as a sol. Cytoplasm in a sol condition can flow freely. ³Next, cytoplasm flows into the cell membrane, extending it to form a pseudopodium. ⁴Then this cytoplasm changes back to a gel. ⁵Finally, other areas of the cytoplasm undergo the same sol-gel transformation forming new pseudopodia. New pseudopodia are formed and others disappear. In this way, the shape of amoebae changes constantly. See Figure 13-1.

Figure 13-1. **Movement of the Amoeba**

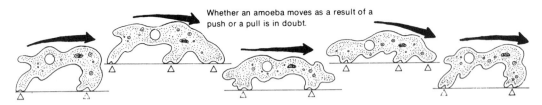

Whether an amoeba moves as a result of a push or a pull is in doubt.

From: Raymond F. Oram, Paul J. Hummer, Jr., and Robert C. Smoot, **Biology: Living Systems** (Columbus, Ohio: Charles Merrill Pub. Co., 1976), p. 327. Reprinted by permission of the publisher.

What?
ANS *Food-getting structures.* Pseudopodia are used in getting food as well as in moving about. Pseudopodia engulf particles of food. The food particles enter the cytoplasm where they are digested. You can see the food vacuoles in which digestion takes place. See Figure 13-2.

Figure 13-2. **Food-Getting Structures of the Amoeba**

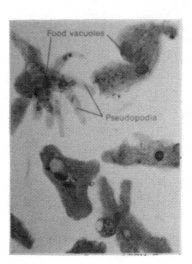

What? ANS *Habitat.* The kind of amoeba studied in the laboratory is commonly found <u>in ponds and other bodies of water.</u> This type of amoeba is harmless. Some amoebae, however, are found in contaminated water. They can cause disease and are quite dangerous to humans.

Skill 4, Exercise 3

<u>Mendel's Initial Experiment</u>

What?
ANS   ***<u>Purpose.</u>*** Mendel experimented with pea plants <u>to discover how traits are transmitted from parent to</u>
        <u>offspring</u>. He wanted to arrive at a set of rules for the transmission of traits. ⟨He had already made⟩
        ⟨certain observations about pea plants.⟩ First, he had seen that traits or visible characteristics are
DEF   hereditary. *<u>Hereditary</u>* means <u>that traits are passed on from generation to generation.</u> Second, he had
        seen that many traits exist in one of two possible forms. For example, pea seeds are either round or
        wrinkled. The stems of pea plants are either tall or short.

What?
ANS   ***<u>Plants used in the experiment.</u>*** For his experiment, <u>Mendel used pure pea plants</u>. *<u>Pure plants</u>* are
DEF   <u>those which after many generations of offspring have the same features as the parents</u>. Pea plants
        reproduce sexually. There are both male and female sex organs in the same flower. Normally, male
        gametes (sex cells or pollen) fertilize eggs of the same flower. Pollen from the anther fertilizes an egg
        in an ovule. See Figure 11-1. The ovule becomes the seed. The ovary develops into a pod which
        protects the seeds. After many generations this process results in offspring with the same features as
        the parents or pure pea plants.

Figure 11-1. **<u>Sexual Reproduction of Pea Plants</u>**

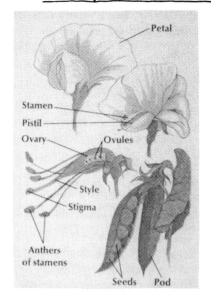

From: Raymond F. Oram, Paul J. Hummer, Jr., and Robert C. Smoot, **Biology:
Living Systems** (Columbus, Ohio: Charles Merrill Pub. Co., 1976), p. 146.
Reprinted by permission of the publisher.

What? *Mendel's accomplishment.* Mendel crossed pure plants having a particular trait with pure plants

ANS having the opposite trait. More specifically, he crossed pea plants which produced round seeds with pea plants which produced wrinkled seeds. To do this, he transferred the pollen of a plant with one trait to plants with the opposite trait, or opposite form of the trait. Pollen was obtained from round-seeded plants and transferred to wrinkle-seeded plants. The opposite procedure was also performed.

What? *Results.* Mendel found that in every case the parental cross (P) yielded offspring with round seeds

ANS
DEF only. The offspring of a parental cross are called the *first filial*, or F_1, *generation*. There were no wrinkle-seeded plants in the F_1 generation.

What? *Law of dominance.* Mendel concluded that for each trait there is one form which "dominates" the

DEF other. A *dominant trait* is the trait which appears exclusively in the F_1 generation. For example, the trait for round seeds is the dominant trait. It "dominates" the trait for wrinkled seeds. The trait which

DEF disappears in the F_1 generation is called a *recessive trait*. In this case, the trait for wrinkled seeds is the recessive trait. Mendel generalized this result and formulated the law of dominance. The Law of

ANS Dominance states that one trait, the dominant trait, dominates or prevents the expression of the recessive trait.

Skill 4, Exercise 4

Acceleration

What? *Acceleration response.* Every driver knows that he must press down the accelerator pedal to increase the speed of his car. All cars do not, however, respond in the same way to a depression of an accelerator. Some will change speed much faster than others. The *acceleration response* refers to the time it

ANS takes to achieve a certain speed starting from rest. If you compare the acceleration of two cars, the one reaching a specified speed in the least time has the greater acceleration. Acceleration can be expressed by the formula:

ANS
$$acceleration = \frac{change\ in\ speed}{time\ required\ for\ change\ to\ occur}$$

What? *Direction and acceleration.* Direction can be part of acceleration. Acceleration then refers to a

ANS change in velocity rather than speed. You can say then that an object can be accelerated by changing either (or both) the speed and direction.

Consider an example. A car travels at a constant speed of 50 miles/hour. It is traveling on a circular track. The car is accelerating because its direction is constantly changing. See Figure 2-1.

Figure 2-1. **Direction and Acceleration**

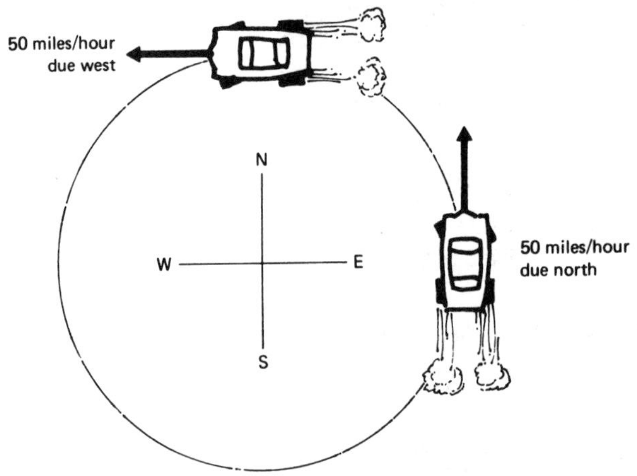

From: Joseph Priest, **Energy for a Technical Society**, ©1975, Addison-Wesley
Pub. Co., Inc., Reading, Massachusettes, p. 50. Reprinted with permission.

Skill 4, Exercise 5

Newton's Second Law of Motion

What? *What causes resistance?* Suppose that you kicked different kinds of objects on the floor with the same
force. Would all of the objects achieve the same acceleration? Surely they would not. Every object
ANS offers some resistance to being put into motion. Resistance is caused by the property of mass known
DEF as inertia. *Inertia* is the property or tendency of a body to resist any change in its state of motion, be it
starting, stopping, or changing it from a straight-line path. The greater the weight, the greater the
mass, the greater the inertia, the greater is the resistance to motion.

How? *Assign a number to measure mass.* A simple balance can be used to assign a number to measure
ANS mass. It compares the unknown mass with the known mass. See Figure 4-1. The units used are
kilograms. The symbol for kilogram is kg.

Figure 4-1. **Measuring Mass**

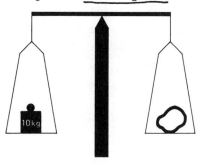

From: Joseph Priest, **Energy for a Technical Society**, ©1975, Addison-Wesley
Pub. Co., Inc., Reading, Massachusettes, p. 31. Reprinted with permission.

What ?
ANS
DEF

ANS

Second law, force, and equations. Newton's second law quantifies these ideas discussed above. The
second law states: "Acceleration experienced by an object is proportional to the net force acting on it
and is inversely proportional to its mass." It then defines *force* as the amount of acceleration given a
mass. In equation form you can say:

$$\text{Acceleration} = \frac{\text{net force}}{\text{mass}}$$

$$\mathbf{a = F/m} \quad \text{or} \quad \mathbf{F = ma}$$

The smaller the mass of an object acted upon by a given force, the greater its acceleration and hence,
final velocity. See Figure 4-2.

Figure 4-2. **Mass and Acceleration**

When the same force is applied to bodies of different masses, the resulting accelerations are inversely proportional to the masses. Succesive positions of blocks of
mass, **m, 2m,** and **3m,** are shown at 1-s intervals while identical forces of **F** are
applied.

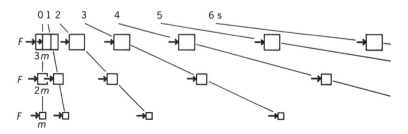

From: Arthur Beiser, **Physics**, 2nd Edition (Menlo Park: CA: The Benjamin/
Cummings Pub. Co., 1978), p. 67. Reprinted by permission of the publisher.

What ?   *The Newton.* It is convenient to use a special unit for force. In the SI system of units (Systeme
International Version of the Metric System), the unit used for force is the newton. Newton is abbrevi-
ANS-DEF ated: N. A *newton (N)* is that force which, when applied to a 1 kg mass, gives it an acceleration of 1
m/s^2. Here, as in Figure 4-2, "s" refers to seconds and "**m**" refers to mass.

Skill 4, Exercise 6

Work from the Physics Standpoint

What ?   *Work and its equation.* From the physics standpoint, physical effort directed toward the production of
something is supplied by a force. Work is said to be done when a force acts on a mass while it moves
through some distance. A physical quantity called *work* can be defined as follows:

ANS-DEF
The work done by a force acting on an object which moves in the same
direction as the force is equal to the magnitude of the force multipled by
the distance through which the force acts.

The definition of work in equation form reads:

ANS

Work = force x distance
W = Fs

How? *Measuring work.* Work can be measured in units of foot pounds and joules. The foot pound (ft. lb.) is
ANS the unit of work in the British system of units. One *foot pound* is equal to the work done by a force of
DEF one pound acting through a distance of one foot.

For example, a man may push a refrigerator. The amount of work he does is equal to the magnitude or amount of force he applies times the distance he moves the refrigerator. He may push the refrigerator 10 feet with a force of 20 pounds. If he does this, he does 20 lb. x 10 ft. = 200 foot pounds of work.

The joule (J) is the unit of work in a more recent system of units, the Systeme International (SI), the
DEF current version of the metric system. One *joule* is equal to the work done by a force of 1N (newton)
DEF acting through a distance of 1m(meter), i.e. 1J = 1N x m. A *newton* is a metric unit of force that is
equivalent to a force of about 1/4 pound.

It's important to realize, however, that a force may be exerted and yet no work may be done. This happens if the force does not act through a distance. For example, a man pushing against a brick wall does no work on the wall if the wall does not move. The product of force and distance is zero because the distance is zero. Only a force that gives rise to motion is doing work.

What? *Positive work and negative work.* It is important to recognize that a force can act in one of two ways.
ANS First, it can act in the same direction as the mass is moving (positive work). Second, force can act
opposite to the direction in which the mass is moving (negative work). In both cases work is being
done. (See Figure 7-1, parts a and b)

Figure 7-1 **Positive Work and Negative Work**

(a)   Force acts in the same direction that the car
moves. This tends to increase the car's speed.
This is a case of positive work.

(b)   Force acts in a direction opposite to that in
which the car moves. This tends to decrease
its speed. This is a case of negative work.

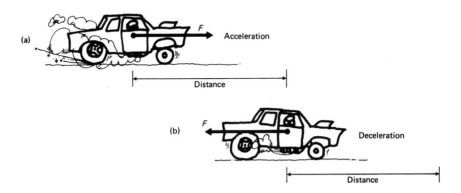

From: Joseph Priest, **Energy for a Technical Society**, ©1975, Addison-Wesley
Pub. Co., Inc., Reading, Massachusettes, p. 34. Reprinted with permission.

What?   *Net work.* The *net work* on an object is the algebraic sum of the positive and negative works done by
ANS-DEF each force acting on the object. Suppose that two forces of +100 N and -50N caused an object to be
moved a distance of 2 meters. Then the work done to the two forces would be:

$$(+100) \ (+2) \ = \ +200J$$
and
$$(-50) \ (+2) \ = \ -100J$$

The net work would be +200 - 100 = +100J

Skill 4, Exercise 7

Dalton's Assumptions

What?
ANS
DEF
DEF
Assumptions. Dalton made two assumptions about atoms and molecules. First, Dalton hypothesized that all the atoms of a particular element are alike and have the same mass. An *atom* is the smallest particle of an element. An *element* is a substance composed of only one type of element; it cannot be broken into other substances by chemical means. Second, Dalton assumed that chemical combination involves chemical bonding. This bonding of atoms of each of the combining elements forms molecules of the resulting compound. Bonding occurs in a predictable manner.

What?
ANS
Law of Conservation of Mass. Dalton's atomic theory helped explain the Law of Conservation of Mass. The Law of Conservation of Mass tells us that there is no detectable gain or loss of mass during any physical or chemical change. Note that the Law of Conservation of Mass is applied to chemical reactions as well as to physical transformations. Atoms involved in both kinds of processes keep their identities. In a chemical reaction, the molecules of the products contain the same atoms that were initially present in the molecules of the reactants. When charcoal burns, for example, there are two reactants. One is an atom of carbon. The second is a molecule of oxygen. One atom of carbon combines with one molecule of oxygen. The product or result is a carbon dioxide molecule. Each individual atom retains its original mass. The sum of masses of the molecules of the product is the sum of the masses of the molecules of the reactants.

Skill 4, Exercise 8

Groups

What?
ANS
Vertical columns. The vertical columns in the table are chemical groups. The elements in any one column therefore make up a group.

What?
ANS
Elements in a group or column. Elements in a group are a family of elements having similar properties. They're more like each other than they're like elements of other groups.

Additional points can be made about elements in the groups which are designated by Roman numerals in two series, A and B. First, the physical and chemical properties of elements in a group quite regularly change in qualities with increasing atomic number. For example, the melting points (Celsius) of the alkali metals are $_3$Li, 180°; $_{11}$Na, 98°; $_{19}$K, 64 °; $_{37}$Rb, 39°; $_{55}$Cs, 29°. See Figure 5-2.

Figure 5-2. **Melting Points of the Alkali Metals**

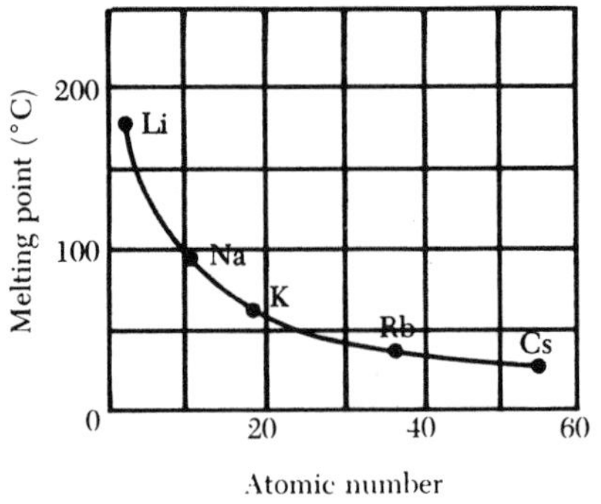

From: Lawrence P. Eblin, **The Elements of Chemistry**, 2nd Edition (Orlando: Harcourt Brace Jovanovich, 1970), p. 67. Reprinted by permission of the publisher.

Second, elements in the same group form compounds with similar formulas. For example, formulas for oxides of Group VB elements are: V_2O_5, Nb_2O_5, Ta_2O_5. Third, in the A groups, metallic properties are accentuated with an increase in the atomic number. Fourth, in the A group, nonmetallic properties diminish as the atomic number increases. Fifth, the first member of a group is often more different from the other members of the group than the other members are different from each other. For example, fluorine does not resemble chlorine as much as chlorine resembles bromine. Also boron is quite different from the other elements of Group IIIA. It is the only nonmetal in the group. Sixth, the members of an A group somewhat resemble members of the corresponding B group. For example, consider manganese in Group VIIB and chlorine in Group VIIA. They both form acids with similar formulas and properties.

What?
ANS

Group names. Certain groups have been given names. Group IA metals (excluding hydrogen) are called alkali metals. Group IIA metals are the alkaline earth metals. On the other side of the table, Group VIIA are the halogens. Group VIIIA are the noble gases.

What?
ANS

What location tells us. Where an element is located in the table tells us its physical and chemical properties. The elements on the left side, for example, are strongly metallic. The elements on the right side are nonmetallic. Then, some elements are between the metals and nonmetals. They are sometimes called metalloids. For example, consider arsenic and selenium. They're sold under the labels, "arsenic metal" and "selenium metal." Chemists would agree, however, that these elements should be classified as nonmetals. Remember also that what is true about one member of a group is usually true about the others in the same group.

Skill 4, Exercise 9

Early Experimentation

What?
ANS

Experiment. Sir Robert Boyle tried to discover if the volume and pressure of a gas were related. He used an apparatus with gas and mercury to measure the relationship. See Figure 7-1.

Figure 7-1. **Apparatus for Determining Relationship of Volume and Pressure of Gases**

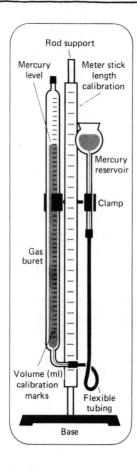

From: Richard M. McGurdy, **Qualities and Quantities: Preparation for College Chemistry** (Orlando: Harcourt Brace Jovanovich, 1975), p. 65. Reprinted by permission of the publisher.

A sample of gas (e.g., air, nitrogen, or carbon dioxide) was trapped in the apparatus. It was trapped over mercury in a closed device used to measure volume. The mercury acted as a piston to change the volume of the gas when the mercury reservoir was raised or lowered. The surface level of the mercury was related to the amount of pressure exerted.

What?
ANS

Result. Boyle found that for gases, as the pressure increased the volume was decreased. Volume was decreased in proportion to the amount of pressure increase. Using a constant amount of gas at a constant temperature, he was able to show this. Look at the figures in Table 7-1. Notice the definition of torr in Table 7-1 also.

Table 7-1. **Pressure versus Volume (at constant temperature)**

Pressure (torr)	Volume (ml)	PV Product
950	30.3	2.88×10^4
750	38.4	2.88×10^4
500	57.6	2.88×10^4
350	82.3	2.88×10^4
300	96.0	2.88×10^4

DEF

A *torr* is the pressure exerted by a column of mercury one millimeter high.

From: Richard M. McGurdy, **Qualities and Quantities: Preparation for College Chemistry** (Orlando: Harcourt Brace Jovanovich, 1975), p. 65. Reprinted by permission of the publisher.

These figures can be placed on a graph. The curve in the graph shows an inverse relationship. As pressure increases, volume decreases. Look at the solid line in Figure 7-2.

Figure 7-2. **Pressure and Volume of Gases**

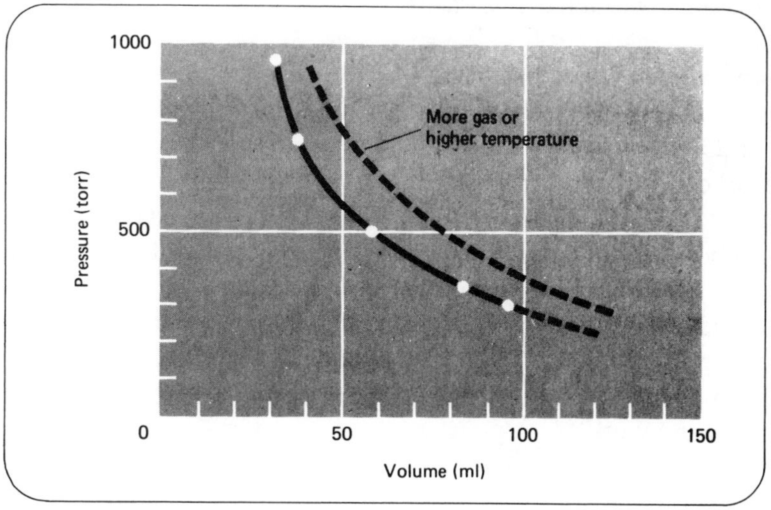

From: Richard M. McGurdy, **Qualities and Quantities: Preparation for College Chemistry** (Orlando: Harcourt Brace Jovanovich, 1975), p. 66. Reprinted by permission of the publisher.

What?

ANS

Equation for relationship. The equation for the relationship between volume and pressure is $\underline{V = k/P}$. V is volume. P is pressure. k is a proportionality constant. This constant, k, depends on the temperature and the amount of gas. As the pressure increases, the volume decreases. The fraction k/P gets smaller. It never becomes zero or negative, however. As the pressure decreases the volume increases. Volume becomes very great as the pressure nears zero.

A sample with more gas or higher temperature could be used too. It would result in a similar curve (dashed line) further from the axis. See Figure 7-2.

Skill 5, Exercise 1
  1.   Sexual Reproduction of Pea Plants
  2.   petal
  3.   ovary
  4.   stigma

Skill 5, Exercise 2
  5.   Forces on Two Colliding Cars
  6.   The auto that is struck is moved forward. At the same time it exerts an equal but oppositely directed force on the moving auto; it is, therefore, pushed backwards.
  7.   It would also be forced to move forward. The middle car would then have both forces acting on it (forward and backward).

Skill 5, Exercise 3
  8.   Gravity and Work
  9.   W
  10.   m
  11.   gravity
  12.   No work is done by gravity

Skill 5, Exercise 4
  13.   Molecules of Compounds
  14.   Molecules of compounds contain different kinds of atoms
  15.   A water molecule consists of two hydrogen atoms and one oxygen atom.
  16.   a hydrogen chloride molecule

Skill 5, Exercise 5
  17.   The Periodic Table of Elements
  18.   Groups
  19.   Periods
  20.   an element's chemical symbol and its atomic number

Skill 5, Exercise 6
  21.   Melting Points of the Alkali Metals
  22.   melting point (C)
  23.   Atomic number
  24.   Rb
  25.   100

Skill 5, Exercise 7
  26.   Apparatus for Determining Relationship of Volume and Pressure of Gases
  27.   flexible tubing
  28.   To support the apparatus and hold the meter stick
  29.   ml (milliliters)

Skill 5, Exercise 8

30. Pressure versus Volume (at constant temperature)
31. the pressure exerted by a column of mercury one millimeter high
32. Volume (ml) and PV Product
33. 38.4 ml
34. They remain the same.
35. It decreases.

Skill 5, Exercise 9

36. Pressure and Volume of Gases
37. volume (ml)
38. pressure (torr)
39. It increases.
40. 500 torr

Skill 6, Exercise 1

I. What Are Algae Like?
  A. Four general traits
    1. possess chlorophyll and carry on photosynthesis
       a. chlorophyll - DEF - green pigments needed for metabolic processes in plants
       b. photosynthesis - DEF - process by which light energy is absorbed and then converted to chemical bond energy of glucose
    2. can live in both fresh and salt water
    3. simple organisms, lacking specialized tissues for conducting water and plant organs such as roots, stems and leaves
    4. have complex reproduction processes
  B. Importance to aquatic communities
    1. primary food producers for other forms of aquatic life
    2. important source of oxygen for them

Skill 6, Exercise 2

    III.   Ciliates
        A. General description
            1. one-celled protozoa (like amoebae)
            2. definite shape with stiff covering bordering their
               cells (unlike amoebae)
            3. on surface have hair-like cilia
               a. cilia - DEF - hair-like structures used for getting
                   food and for moving about
        B. Paramecium
            1. complex organism in which different parts of cytoplasm
               do different jobs

See Fig. 13-3 2. 2 types of nuclei within single cell
pg. 221   a. macronucleus which controls basic activities
               b. micronecleus involved in reproduction
            3. moves by beating cilia on its surface

See Fig. 13-4   a. glides rapidly through water
pg. 222   b. rotates as it moves

Skill 6, Exercise 3

    II.   Further Experimentation
        A. F_2 generation
            1. second filial or F_2 generation - DEF - offspring of F_1
               generation cross
        B. Results
            1. ratio of 3:1; 3 round seeded to each wrinkle-seeded
               a. of 7,324 offspring, 5,474 had round seeds and 1,850
                   had wrinkled seeds
               b. additional experiments had same 3:1 ratio in F_2 generation

Skill 6, Exercise 4

    I.   Speed
        A. Speed and its formula
            1. speed - DEF - distance traveled in some period of time
            2. formula:

$$speed = \frac{distance\ traveled}{time\ period} \quad or \quad s = \frac{d}{t}$$

                a. example: runner completes 100 yd. dash in 10 sec.

$$\frac{100\ yds.}{10\ sec.} = \frac{10\ yds.}{1\ sec.}$$

        B. Distance
            1. formula: distance = speed x time or d=st
                a. example: car travels at 60 mph for 3 hours.
                    d = 60 mph x 3 hours = 180 miles

Skill 6, Exercise 5

    I.   Newton's First Law of Motion
        A. First law
            1. "Every body continues in its state of rest, or of uniform
                motion, in a straight line unless it is compelled to change
                that state by forces impressed on it."
                a. uniform = constant
                b. motion = velocity
        B. What this law suggests
            1. looking for force that causes acceleration
                a. often obvious as when baseball is batted
                b. other times not so obvious as when gravity accelerates
                    ball dropped from window

Skill 6, Exercise 6

    II.   Work Done Against Gravity
        A. Acceleration due to gravity
            1. when gravity is only force, acceleration is same for all masses
                a. example: velocity changes by 9.8 meters per second for every second that object falls
                b. m/sec^2 - abbreviation for meters per second per second
            2. symbol "g" stands for this acceleration
            3. $g = 9.8 \ m/sec^2$
        B. Weight and its symbols
            1. force of gravity on an object is same as its weight
                a. weight - DEF - mass multiplied by acceleration due to gravity
                b. symbols: $w = mg$
        C. Computing work done against gravity
            1. 2 points
                a. $w = mg$

See Fig. 7-2   b. in order to raise object to height above its original position,
pg. 240   a force of mg must be exerted on it
            2. equation: $W = mgh$
        D. Importance of route taken
See Fig. 7-3   1. only height is involved
pg. 241   2. route not significant

Skill 6, Exercise 7

    II.   Molecules
        A. Molecule defined
            1. molecule - DEF - smallest particle of substance as it normally exists
            2. 3 additional points
                a. minute, uncharged particle
                b. moves as whole
                c. component parts not detached during motion
        B. Molecules of compounds and elements
See Fig.3-1   1. molecule of compound consists of more than 1 kind
pg. 244   of atom
See Fig. 3-2 2. molecule of element consists of 2 or more atoms of
pg. 244   same kind

Skill 6, Exercise 8

I.   Rows
    A. Horizontal rows
        1. chemical periods
            a. elements in any 1 row make up series/period
        2. things to note
            a. 7 rows; numbered 1-7
            b. # of elements in each row/period increases in

See Table 5-1   order 2, 8, 8, 18, 18, 32, 17 (last period incomplete)
pg. 248   c. 15 elements from each of 2 long periods printed
                below table
    B. Last element
        1. noble gas
            a. nonatomic/single atom
            b. very little chemical reactivity
            c. examples: helium, neon, argon

Skill 6, Exercise 9

II.   Conclusions
    A. How volume changes
        1. product of new pressure and volume always equals
            product of initial pressure and volume
        2. equation: $P_f V_f = P_i V_i$
            a. subscript f = final
            b. subscript i = initial
    B. Figuring final volume
        1. multiply initial volume by ratio of pressures
        2. equation:
$$V_f = V_i \times \frac{P_i}{P_f}$$
    C. Sample problem and solution
        1. given: sample of gas
            a. room temperature
            b. volume = 35.0 milliliters at pressure of 725 torr
        2. question: what would its volume be at 760 torr?

3. solution: 2 steps
    a. write $P_fV_f = P_iV_i$ using real figures: 35.0 ml x 725 torr = V_f ml x 760 torr
    b. use real figures in the formula for figuring final volume:

$$V_f = V_i \text{ ml } (35.0 \text{ ml}) \times \frac{P_i \ (725 \text{ torr})}{P_f \ (760 \text{ torr})} = 33.4 \text{ ml}$$

Skill 7, Exercise 1   Ecosystem & Energy

  I.   Ecosystem - DEF - interaction of community with its environment
  II.   Ways in which living things within community interact
     A. With own species
     B. With other species
     C. With physical environment
  III. Interactions between living organisms and nonliving things
     A. Give and take relationships
        1. living things take materials and energy from environment
        2. as they use materials and energy they give some materials
           and energy back to environment
        3. energy cannot be completely recycled - must be added
  IV. Source of energy added constantly - sun
     A. Light energy used by organisms in series of energy transfers
        1. first transfer performed by plants - primary producers
           a. use photosynthesis to capture light energy
           b. store it in form of chemical bonds
        2. chemical bond energy passed along to higher level organisms

Skill 7, Exercise 2   The Food Chain

    I.   Food chain involves transfer of materials and energy from organism to organism
        A. Always starts with autotrophs
            1. DEF - organisms that can produce own food (e.g., green plants)
            2. only organisms that can change light energy to chem. bond energy
        B. Series of relationships
            1. plants receive energy from sun and change it to chem. bond energy
            2. plants are eaten by herbivores
                a. DEF - plant-eating animals
                b. chem. bonds broken and reformed into new bonds in animal's cells
            3. animals called carnivores eat herbivores
                a. DEF - meat eaters
                b. breaking and reforming of chem. bonds occur again
        C. 4-step chain — example
            1. plant — producer
            2. caterpillar — herbivore
            3. mockingbird — primary carnivore
            4. hawk — top carnivore
    II. Food chains interlocked in complex pattern of feeding relationships called food web
        A. Many species found at each feeding level
        B. Energy transformations occur all along

Skill 7, Exercise 3   Evolution of Species

    I.   Species - DEF - group of organisms which normally interbreed in nature
        A. A closed gene pool into which "foreign" genes cannot enter by normal mating
        B. Result of interbreeding is fertile offspring
    II. Reproductive isolation - gene pool of 1 species separated from all other species
        A. Various factors provide barriers to interbreeding
            1. differences in mating habits
            2. inability of sperms to fertilize eggs
            3. physical impossibility of mating

    B. If different species mated, embryos would not develop or be suited
       to environment

III. For new species to evolve must be division/separation/isolation
    of population
    A. Usually caused by physical feature - geographic isolation
    B. Series of things happen then, encouraging evolution of
       new species
       1. gene pools of 2 isolated groups do not interbreed
       2. each gene pool evolves by natural selection in its environment
       3. 2 groups develop their own set of traits
          a. classified as subspecies because of different traits
       4. if isolation continues and further changes evolve, 2 groups may
         be unable to interbreed
          a. 2 groups now separate species
          b. evolution of new species has occurred

Skill 7, Exercise 4   Friction

  I. How it is different from inertia
    A. Inertia - DEF - tendency of bodies to maintain their original states
       of rest/motion in absence of net forces acting on them
    B. Friction - DEF - actual force that comes into being to oppose motion
       between 2 surfaces which are in contact
  II. Friction sometimes desirable, other times not
    A. Desirable in many instances
       1. when you use nails, bolts, screws, fastening action result of friction
       2. walking
    B. Not desirable
       1. when reduces efficiency
          a. e.g. lubricants and special devices used to reduce it in industry
III. Frictional forces when try to move box across level floor
    A. Box is stationary; no horizontal forces act on it
    B. As we begin to push, box remains in place; floor exerts force
       causing frictional force opposing force we apply
    C. As we push harder, frictional force increases to match our
       efforts; box remains stationary
    D. We exceed frictional force and move box
       1. frictional resistive force decreases, then remains constant
       2. net force on box = force of push - force of friction

Skill 7, Exercise 5   Centripetal Force

I.   Centripetal force - DEF - force seeking center, needed for circular motion
     A. Equals inward force on object moving in curved path
II. When object travels in circle, direction of velocity changes constantly
     A. Changing velocity = acceleration
     B. Object must be acted upon by force
        1. force must be directed toward center of circle since object's path
           is circle - centripetal force
III. Centripetal force acts whenever rotational motion occurs
     A. Provided in various ways
        1. gravitation keeps planets moving around sun, moon around earth
        2. friction between car's tires and road needed for car to round curve
IV. How great a centripetal force needed to keep given object moving in
     circle at certain speed
     A. Determine centripetal acceleration
        1. involves change of direction causing object to change its motion
           from straight line into circular path - always directed toward center
           of path
        2. formula: orbital velocity of object moving in circle squared over
           radius of circle, V^2/r

Skill 7, Exercise 6   Energy

I.   Energy - DEF - that property whose possession enables something to perform work
II. 3 broad categories
     A. Kinetic energy
         1. DEF - energy something possesses by virtue of its motion
             a. e.g., fast-moving car will do more work on phone pole than slow-moving car of same type
     B. Potential energy
         1. DEF - energy something possesses by virtue of its position
             a. car at top of hill has potential energy; if it rolls down hill, potential energy converted to kinetic energy
             b. gasoline in tank has chemical energy/potential energy; when ignited, heat energy is released and chem. energy becomes kinetic energy
                 (1) this is molecular kinetic energy - DEF - kinetic energy released only by appropriate chem. or phys. reaction
     C. Rest energy
         1. DEF - energy something posseses by virtue of its mass
         2. e.g., in mass of nucleus of atom; when released, mass is converted to kinetic energy

Skill 7, Exercise 7   The Concept of Mass

    I.   Mass - DEF - a measure of amount of matter present
   II.   several forms
       A. Liquid
       B. Gas
       C. Solid
  III.   Can be determined in 2 ways
       A. Weighing - using spring scale to measure gravitational attractive
          force between object and earth
          1. Magnitude of force and likewise weight of object varies,
             depending on 3 factors
             a. mass of object
             b. mass of earth
             c. distance between center of object and
               center of earth
       B. Using chemical balance in lab
          1. specimen's attraction to earth is balanced against attraction
             earth has for known set of masses
          2. doesn't vary like weight
  IV.   Density can be figured
       A. DEF - mass per unit volume
       B. 3 steps
          1. specimen weighed
          2. volume measured
          3. mass divided by volume to get density

Skill 8, Exercise 1
1. e
2. b

Skill 8, Exercise 2
3. F
4. T
5. F

Skill 8, Exercise 3
6. d 1.
   b 2.
   f 3.
   a 4.

Skill 8, Exercise 4
7. speed
8. d = s (speed) x t (time)
9. its speed and direction in which it is moving

Skill 8, Exercise 5
10. No. The cell structure of blue-green algae and their method of reproduction can be described as follows.
11. Yes
12. Yes
13. Ask your instructor to read your answer.

Skill 8, Exercise 6
14. Yes
15. Yes
16. Yes
17. Ask your instructor to read your answer.

Skill 8, Exercise 7
18. c
19. a

Skill 8, Exercise 8
20. F
21. F
22. T

Skill 8, Exercise 9
23. g 1.
    d 2.
    a 3.
    c 4.
    f 5.

Skill 8, Exercise 10
24. force; distance
25. meter
26. force of about 1/4 pound

Skill 8, Exercise 11
27. Yes
28. Yes
29. Yes
30. Ask your instructor to read your answer.

Skill 8, Exercise 12
31. No. The methods used to figure speed and to figure distance are quite different.
32. Yes
33. Yes
34. Ask your instructor to read your answer.

Skill 8, Exercise 13
35. No. It is often important to know the velocity of a moving body.
36. No
37. No
38. Ask your instructor to read your answer

Skill 9, Exercise 1
1. QC
   21.2
   M366
   Principles of Physics
   Jerry B. Marion and
       William F. Hornyak
   Saunders College Pub.
   1984

2.   author card
3.   Philadelphia
4.   772 pages
5.   illustrations, index
6.   Physics

Skill 9, Exercise 2

7.   QD
     31.2
     M388
     General Chemistry
     Donald A. McQuarrie and
          Peter A. Rock
     W.H. Freeman
     1984
8.   subject card
9.   New York
10. 1063 pages
11. illustrations, index
12. Chemistry

Skill 9, Exercise 3

13. QH
     308.2
     R34
     **Biology: The Science
          of Life**, 2nd ed.
     Joan E. Rahn
     Macmillan
     1980
14. subject card
15. New York
16. 673 pages
17. Plates, illustrations, biblio-
          graphies, and index
18. Biology

Skill 9, Exercise 4

19. QC
     23
     S458

**Physics, for Scientists
     and Engineers**
Raymond A. Serway
Saunders College Publ.
1982

20. title card
21. Philadelphia
22. 883 pages
23. illustrations, bibliographical
          references, and index
24. Physics

Skill 9, Exercise 5

25. QD
     31.2
     E22
     General Chemistry
     Darrell D. Ebbing
     Houghton Mifflin Co.
     1984
26. subject card
27. Boston
28. 970 pages
29. plates, illustrations, and index
30. Chemistry

Skill 9, Exercise 6

31. Plants
32. Disease and pest resistance
33. "Cultured Cells of White Pine
          Show Genetic Resistance to
          Axenic Blister Rust Hyphae"
34. A.M. Diner and others
35. bibliography, illustrations
36. **Science**
37. Volume 224
38. pages 407-408
39. colon (:)
40. April 27, 1984

Skill 9, Exercise 7

41. Chemicals
42. Safety Measures
43. "Chemistry Workshops to Prolong the Lives of Your Favorite Janitors"
44. J.N. Aronson
45. none
46. **J Chem Educ (Journal of Chemical Education)**
47. Volume 60
48. pages 1036-37
49. colon (:)
50. December 1983

Skill 9, Exercise 8

51. Hydrogen
52. Spectra and spectroscopy
53. "Observations of Lyman — Emissions of Hydrogen and Deuterium"
54. J.L Bertaux and others
55. bibliographical footnotes, illustrations
56. **Science**
57. Volume 225
58. pages 174-76
59. colon (:)
60. July 13, 1984

Skill 9, Exercise 9

61. Biology
62. Classification
63. "Taxonomy: What's in a Name?"
64. C.J. Cole
65. illustrations
66. **Nat Hist (Natural History)**
67. Volume 93
68. pages 30+
69. colon (:)
70. September 1984

Skill 9, Exercise 10

71. Heredity of diseases
72. none
73. "Gene Therapy Method Shows Promise"
74. G. Kolata
75. none
76. **Science**
77. Volume 223
78. pages 1376+
79. colon (:)
80. March 30, 1984

✎ Attitude

✎ Motivation

✎ Anxiety

✎ Concentration

Notes

Notes

Bogue, Carole

AUTHOR

Studying in the Content Areas

TITLE 2nd Ed.

DATE DUE	BORROWER'S NAME